About Island Press

Since 1984, the nonprofit organization Island Press has been stimulating, shaping, and communicating ideas that are essential for solving environmental problems worldwide. With more than 1,000 titles in print and some 30 new releases each year, we are the nation's leading publisher on environmental issues. We identify innovative thinkers and emerging trends in the environmental field. We work with world-renowned experts and authors to develop cross-disciplinary solutions to environmental challenges.

Island Press designs and executes educational campaigns, in conjunction with our authors, to communicate their critical messages in print, in person, and online using the latest technologies, innovative programs, and the media. Our goal is to reach targeted audiences—scientists, policy makers, environmental advocates, urban planners, the media, and concerned citizens—with information that can be used to create the framework for long-term ecological health and human well-being.

Island Press gratefully acknowledges major support from The Bobolink Foundation, Caldera Foundation, The Curtis and Edith Munson Foundation, The Forrest C. and Frances H. Lattner Foundation, The JPB Foundation, The Kresge Foundation, The Summit Charitable Foundation, Inc., and many other generous organizations and individuals.

The opinions expressed in this book are those of the author(s) and do not necessarily reflect the views of our supporters.

THE JEWEL BOX

The Jewel Box

How Moths Illuminate Nature's Hidden Rules

Tim Blackburn

ISLANDPRESS | Washington | Covelo

Library of Congress Control Number: 2022946106

Manufactured in the United States of America
10 9 8 7 6 5 4 3 2 1

Keywords: Acer Sober, biodiversity, biology, biophilia, Box-tree Moth,
Codling Moth, communities, competition, ecology, Charles Darwin, Devon,
Dingy Footman, extinction, Goat Moth, Gypsy Moth, Hexapoda, insects,
Lepidoptera, London, macroecology, macromoths, micromoths, moths, moth
trap, niche theory, Oak Eggar, Poplar Hawk-moth, populations, Silver Y,
speciation, species, species richness, Stout Dart, Uncertain

For Milly
More precious than jewels

Contents

Introduction

The Moth Trap

Nature is painting for us, day after day, pictures of infinite beauty if only we have the eyes to see them.
— John Ruskin

The moth trap, on holiday in Devon.

I n July of 2018, for my fifty-second birthday, my wife bought me a black plastic box, two Plexiglass sheets, and a light fixture with a 20-watt fluorescent bulb. The box came flat-packed, which meant that I had to draw on my limited practical skills to intersect the tabs with their corresponding slots, resulting in an open-topped cube about fifty centimeters on each vertex. I screwed the electrics to a bar that ran across the top, setting the socket above the box but sheltered under a wide, white plastic disk. The light bulb bayoneted into the socket, and hung down. The Perspex sheets slotted at a forty-five-degree angle from the rim, forming a wide, transparent slide down into the interior, ending at an opening the size and shape of a letterbox.

A black plastic box with a light on top might seem like an odd choice for a birthday present, but it was what I wanted. It sounds dourly functional. But it is also a box of enchantment, one that can conjure life out of thin air. I put it outside that July evening as dusk was beginning to fall, plugged it in, and watched the glow from the light start slowly to build. Then I went to bed, already excited, hoping the conjuring trick would work. I woke early and expectantly the next morning, and went outside to see. The box had indeed performed its magic, and there, inside, was the reveal—a scattering of jewels. A moth trap, festooned with moths.

I'd first got into moth trapping while leading undergraduate field courses to the Kindrogan FSC (Field Studies Council) center in Scotland (now sadly closed to such trips). As a university academic, one of the joys of my job is that I get to interact with young people eager to learn. Yet, it's not only student minds that come away from field courses changed—they can be a revelatory experience for teacher and taught alike. Even students who are interested in biodiversity aren't nearly close to knowing all the different forms of life with which they share their surroundings.

We always set pitfall traps—plastic cups sunk into the soil like golf holes—to catch ground-dwelling invertebrates, and every year I have to encourage the students not just to pick through the harvest with the naked eye, but also to use a dissecting microscope, and to turn the magnification up high. Without this, they wouldn't see the tiny springtails that inevitably fall in. These animals are in the order Collembola, and are among the closest relatives of the insects; indeed, when I was a student, they were considered to *be* insects. Taxonomic orthodoxy has changed since then, and now they tend to be classified as the earliest divergent branch from the part of the tree of life that is the subphylum Hexapoda, of which insects are the most familiar manifestation. Like insects, springtails have three body sections—head, thorax, abdomen—with six legs attached to the thorax. *Unlike* insects, their mouthparts are hidden inside the head capsule. Most springtails have an organ known as the furcula folded underneath the abdomen, which is responsible for their name. When triggered, this sprung fork can catapult the springtail into the air and away from its enemies.

The springtails caught at Kindrogan pack all of this structure into an animal barely a couple of millimeters (a tenth of an inch) in length. This miniaturization, combined with their unassuming habit of hiding in the leaf litter and soil, means that most students don't know that these organisms even exist. The feeling of offering someone the revelation of an animal they'd never even dreamt of never gets old.

There was a moth trap at Kindrogan, and one summer I asked their staff to run it. This one was a Robinson trap—a large, bucket-like container with a ring-shaped lid and a powerful mercury-vapor bulb. Moths are attracted to its light and drop into the container below to await identification and release in the morning. The trap was sited in a purpose-built shed on a grassy bank looking east over the River Ardle. On that first June morning, the students and I were torn between crowding in to examine what the trap had caught, and hanging back for fear of crushing the delicate insects underfoot—many moths are lured to a moth trap but don't quite make it inside. Most species sport the colors of camouflage, and it takes some time to train one's eye to spot them in long grass.

For some moths, greens predominate for a background of summer leaves. Others are mainly brown for wood and stone. The lemon yellow of a Brimstone Moth blends surprisingly well into darker tones of grass, especially given the brown patches that dot the leading edges of its wings to break up its shape. Clouded-bordered Brindles have a similar strategy to disrupt their outline, and their shades of brown render them shadows between the stems. Even the monochromes of the Clouded Border are hard to spot among sparkling drops of dew.

The shed itself was dotted with insects, too. The long white wings of a male Ghost Moth were aligned with the grain of a board. A Coxcomb Prominent sat in a gap between walls and roof, looking like a sycamore key that had become lodged as it fell. A Gold Swift stood out boldly against the gray planking. And then when we opened the trap—a cornucopia inside. Buff-tips, Hawks, Carpets, Waves, and more. I couldn't identify them all—not then—but I could see what a proliferation of species we'd summoned. The trap had given us all a glimpse of unexpected biodiversity in a country I thought I knew well.

Arriving back at King's Cross after a week out with a group of students is always a time of mixed feelings. Home is just a handful of tube stops away, with a weekend of much needed rest, sleep, and family time awaiting. Fully immersive teaching with long days in the field is exhausting. But being back in London is always accompanied by an ineffable feeling of loss.

After a week spent in the pine forest and mountains surrounding Kindrogan, the city is an impossible accretion of brick and people, of air and noise pollution. Senses that have come alive in the highlands have to be dialed back down to cope with the urban environment. The journey from Scotland to London takes one through decades of biodiversity loss in a single day, and feeling that loss takes a toll.

If you live and work in a city, it's easy to feel detached from nature—but it's still there if you look. Hampstead Heath is a short walk from my London flat, and the diversity of life it supports is surprising. It often helps to adopt the viewpoint of a five-year-old, getting down on hands and knees, face close to the ground. Even the areas manicured for sport are home to more than just grass. Broad-leaved Plantain and Knotgrass make a living there, lying too low for the blades of the mower. Just a few feet away, where the Heath rangers leave the field uncut, the grass grows tall enough to flower, resolving the stems into Fescue, Cocksfoot, and Foxtail. Yellow buttons of Creeping and Meadow Buttercup mix with the white of Clover. Common Vetch drapes itself across the grass, and

Creeping Thistles lie low, if not yet flagged by their flowering spikes. Picnics in the long grass here need care. And of course, the plants themselves are al fresco dining for a variety of animals. When it isn't raining, butterflies transect the field, and bumblebees and honeybees make beelines from flower to flower. Caterpillars and grubs chew on the leaves, burrow their way into grass stems and thistle buds, and even mine their way between leaf surfaces. Look up and you may see a bird of prey overhead—a Kestrel frequently hovers over the slope here, but Common Buzzard, Red Kite, Sparrowhawk, Peregrine, and Hobby all hunt over the Heath. An early morning visit might produce a Red Fox trotting away to cover. More likely mammal sightings are Brown Rats, or their arboreal cousins, the Gray Squirrels.

Even a short walk can be hard to snatch, though, against a backdrop of work and family. I miss those opportunities for escape. It was something I was feeling especially keenly at the end of that last trip to Kindrogan when the obvious hit me: I could make nature come to me. Why wait a year to run a moth trap again?

My birthday trap was a lightweight Skinner, entry-level, with an actinic bulb as light source, running off mains electricity. It was sited in our only outside space: thirty feet off the ground, on the roof terrace of our flat in the London Borough of Camden. The terrace overlooks some mature gardens, facing a line of tall limes, with large cherry and pear trees in sight. Yet this is very much an urban location in an area with no shortage of other light sources. Before that first morning, I wondered whether this restricted patch of green would house *any* species of moths, and, if it did, whether any would find their way into the trap. Happily, the answer to both questions was yes.

Experience in science tells us that answers to questions simply lead to more questions: as the sphere of knowledge expands, so does the area of interface between known and unknown. So it was, too, with the moth trap. The most pressing new question that first morning was: What were the identities of all the species that had appeared in the box overnight? This was not a trivial task—there were more than eighty moths to pick out and identify. Giving things names matters. It is how we begin to quantify our experience of the natural world. The moths themselves aren't knowledge yet. The first step is to resolve them into their constituent species.[i]

As we will see, our knowledge of most groups of species in most parts of the world is decidedly poor—we are just beginning to describe the diversity of organisms with which we share the planet. Yet there are some notable exceptions, one of which is British moths. Britain is blessed with some excellent field guides to these insects, beautifully and expertly illustrated with paintings and photographs that depict almost all the species found here. Even with these guides, though, it takes time to get one's eye in. Different species prefer different habitats and fly at different times of year, and these ecological details are not apparent simply from looking at pictures in a book. The field guides have this information, but there are hundreds of pages of text to leaf through. It's a slow process. Fortunately, again, the UK moth trapper is blessed with additional support. Type "What's flying tonight?" into an internet search engine and it will link you to a web page that uses your location and the date to produce a list of the species most often recorded there and then. Each comes with photographs and a calendar bar showing when in the year the adults are on the wing. It's based on millions of moth records logged by the charity Butterfly Conservation through their National Moth Recording Scheme, a wonderful application of accumulated scientific data to the public appreciation of biodiversity. Still stumped, one can go to social media, where experts are happy to help you identify all sorts of animals from photographs; moth novices can ask @MothIDUK for assistance (but please consider a contribution for this service if you can). Even with all this help, though, some moths cannot be specifically named without dissecting their genitals, and must simply be logged as a part of an "agg." to denote that they belong to an aggregate of species that are indistinguishable without the skills of the specialist. I am obsessive about identifying animals, but (so far) I draw the line at killing them to satisfy my obsession.

That first morning on my terrace, I began, slowly, to match species to

---

i. If you're wondering what I mean by a "species," that will become clearer before this chapter is out.

their names. The Dun-bar and Knot Grass. Tree-lichen Beauty, Gypsy, Jersey Tiger. Pale Mottled Willow and Dingy Footman. And most aptly, the Uncertain—most of the moths started with this label, but only two finished with it. (Adults of this species are very similar to others that fly at the same time of year. More of this anon.) Finally, after a good chunk of the morning, I'd assigned eighty-two individual moths to twenty-eight different species.[ii]

All these animals had appeared as if by magic on a small roof terrace in urban London. The entomologist and writer E. O. Wilson, who coined the term *biophilia* to describe the innate affinity people have with the natural world, noted that "Every kid has a bug period. . . . I never grew out of mine." That morning, I grew back into mine.

We have given names to more than a million different animal species, but this is certainly only a fraction of the total. Estimates of global animal species numbers range from three million to one hundred million, depending on the method used to extrapolate from those species that are currently known to science, but the actual number seems most likely to be nearer the lower end of this range. One credible recent study calculates just shy of eight million species. This is phenomenal diversity, when, as far as we know, the presence of even one living species sets our planet apart from all others. Yet, of all animal species so far named, roughly one in ten is a moth—around 140,000 species in the order Lepidoptera. One in nine, really—another 20,000 or so Lepidoptera are butterflies, which we distinguish colloquially, but which are just a subgroup of moths that have taken to flying by day. The true number of moth species worldwide is likely to be far higher—most species are found in tropical forests, which remain poorly explored in comparison to the temperate latitudes in which most scientists and taxonomists live and work. Why then, given all this diversity, had my trap pulled twenty-eight species of moth out of the London undergrowth? What

---

ii. A few more individuals had been lured in but escaped during my clumsy attempts to extract them from the trap.

was it that determined twenty-eight? Was there anything that could be done to increase that number? And what would happen if we tried to bring more species into the local environment—would I end up with a richer catch, or would present incumbents simply be squeezed out? What about if we took some species away—would other species move in, or would our local moth community just be the poorer?

There were twenty-eight species in my trap, but eighty-two individual moths. Almost a third of the catch comprised just two species—the Tree-lichen Beauty was the most numerous, with thirteen individuals, but the Jersey Tiger was close behind with twelve. Add in the Dun-bar (eight), Riband Wave, and Codling Moth (six of each), and five of the twenty-eight species accounted for more than half of the moths caught. Most species were represented by ones or twos. Is that typical?

Could the species I caught live elsewhere—indeed, do they? The two most common species in the trap that morning were ones I'd not seen in Kindrogan, which suggested that something was different between these two locations. The habitats around the trap—Camden and Kindrogan—were obviously quite distinct, but perhaps it was just down to geography—do we expect to get the same set of moths in the trap if we shift it 700 kilometers, or not? And what does it tell us if we do? The different sets of moths in Kindrogan and London could also be just an issue of the time of year—spring and summer come later to Scotland than to southeast England, so maybe the Kindrogan trap would be full of Tree-lichen Beauties and Jersey Tigers come August?

No species is an island, entire of itself. All animals must consume to survive, so the presence of food is important. Moths are holometabolous insects, which means that they develop through a life cycle of egg, larva (the caterpillar stage, which is split into a variable number of instars,[iii] punctuated by shedding of the hard chitinous exoskeleton to create room for expansion of the body and allow onward growth), pupa (or chrysalis, in which the miraculous transformation from caterpillar to

---

iii. Caterpillar skins can only stretch so far as they grow, and every so often they need to molt. It's their equivalent of us moving through shoe and clothing sizes. Each molt takes them up an instar. Some moths need to molt more times than others, generally because they are growing to larger sizes. (See chapter 2 for an example of just how much some caterpillars can grow.)

moth takes place), and finally adult. Most consumption is done by the caterpillar; this must start out being able to fit inside a tiny egg, but then accumulate enough raw material to effect the metamorphosis to an adult capable of laying eggs of its own (up to 20,000 in some cases). Most caterpillars are vegetarians, but their tastes vary enormously. The Lesser Broad-bordered Yellow Underwing has a mouthful of a name, but the list of plants its caterpillars will eat is much longer—from Dead-nettles, Docks and Mayweeds, to Sallow, Hawthorn, and Blackthorn. The Marbled Beauty, on the other hand, develops on lichens, which provide slimmer pickings and do not do well in polluted areas. Both species were on my Inner London roof terrace that first morning of trapping.

Moths are consumers, but are also often the consumed. They are links in the web of life, parts of the habitats they occupy, and habitats themselves, for communities of predators and parasites. The moth trap can illustrate this. It sometimes attracts wasps—the colloquial yellowjackets—especially in autumn, which can bring tension to the morning's proceedings. Yellowjackets are important predators of other insects, doing a largely unheralded service as pest controllers in gardens and crops. They often buzz in to try their luck with the moths resting around the trap. Other wasps are also drawn to it, too—parasitoids that lay eggs in the bodies of caterpillars, hatching to eat the host alive from the inside. Sometimes moths are both consumers and consumed: those Dun-bars I caught grew up as omnivores, feeding on leaves but sometimes also on the caterpillars of other moths. What effects do all of these interactions have on the populations of the moths I was catching? Are moth numbers determined by what they eat, or by what eats them?

There is an extensive and venerable branch of science essentially devoted to these questions. This is the science of ecology, which has been my career for the past three decades. I'm deeply interested in questions relating to the abundance, distribution, and richness of species, which I mainly research using information on a group of animals that has always been my first love—birds. My familiarity with this sort of question, and with birds, has certainly not bred contempt, but often it takes the contrast of new experiences for us suddenly to see the familiar in a different light. We take our surroundings for granted. My moth trap had given me a new perspective.

Ecology was first defined in its modern form—as Öekologie, in his native German—in the 1860s by the biologist Ernst Haeckel. The etymological roots of *ecology* combine the ancient Greek words *oikos*, from which we get "eco-," and *logos*, for the principle of order and knowledge. *Oikos* does not have a single meaning; it refers to the family, the family's property, or the house. Why it prefixes *economy* is clear. For the study of the interactions between organisms and their environment—which is one definition of ecology we might use—"eco" relates to the third meaning, and we often give it a more personal interpretation. Literally, ecology is the study of our home. The relevance and logic of this to an ecologist is obvious: we are organisms, and our environment matters fundamentally to us. Haeckel's formal definition was "the comprehensive science of the relationship of the organism to the environment."[1]

Haeckel was a keen disciple of Charles Darwin, and so it seems appropriate that the definition of ecology has evolved over the years. Although Haeckel identified the essence of the subject, he codified it in a form that was arguably too general, and too vague. What isn't ecology, under his definition? What, exactly, is ecology trying to explain? It was not until almost a century later that we got a clear answer to this second question, thanks to the Australian ecologist Herbert Andrewartha. He revised the definition of ecology to "the scientific study of the distribution and abundance of organisms"[2]—the crux is to try to understand where organisms are found, how many are found there, and why. With minor tweaks (Canadian Charles Krebs advocated for the addition of "the interactions that determine,"[3] for example), this is the definition that most ecologists use today. It identifies why ecology is key to understanding the contents of a moth trap.

A moth trap may be a source of wonder for the biophiliac, but it is also an effective scientific tool. The animals that it conjures out of thin air are samples of the wider moth community in the immediate area, or, in some cases, of the moths that are passing through it. They are a snapshot, a fragment of the wider panorama. By piecing together snapshots we can begin to see the bigger picture.

For moths the number of snapshots available is huge, at least in the UK. These islands are home to an extended community of amateur trappers who write up their nightly catch and submit details to regional or national record schemes. Since 1968, a countrywide network of moth traps has been coordinated by the agricultural research station at Rothamsted in Hertfordshire. Historical records that antedate the existence of these schemes can in some cases be extracted from old notebooks and added to the picture. The high-resolution image that results means that we can start to pick out details, and individual moth trappers can see where the pixels they provide fit into the broader patterns that emerge. These patterns become the basis for ideas about how the natural world works, which we can then put to the test by further observations—or better, by experimentation. Are we seeing a random set of individual animals of a random set of species thrown together by chance—or are there rules? And if there *are* rules, what sort of rules are they? Gradually, our understanding of the world around us improves. But the scale of the task should not be underestimated. The natural world is fiendishly complex.

Imagine that you had the technology that would allow you to scan our planet with such detail that you could map the identity and location of every individual animal, plant, fungus, bacteria, archaean, and virus.[iv] What sort of picture would this give you?

Each of the individuals scanned would belong to a species. On top of the eight million or so animal species, estimates suggest about a further million other eukaryotes (roughly 30 percent plants, 60 percent fungi, and the rest protozoa and algae). In comparison, estimates of prokaryote (bacteria and archaea) species numbers range from the surprisingly low (a minimum of around 10,000) to the surprisingly high (perhaps 1 trillion).[v]

---

iv. These are the major kingdoms across which we currently consider life to be distributed, although the number of kingdoms we consider there to be has increased over the history of biological science, and no doubt will continue to change.

v. This enormous uncertainty is partly because we have trouble understanding what the term *species* means for a prokaryote, given our own eukaryotic perspective (and we

These are numbers of *species*, though. We sometimes know the number of *individuals* for a given species very well, but only when that species is so rare that we are worried for its future—the 209 individuals of Kakapo, the large, flightless New Zealand parrot, for example.[vi] Mostly, we have to estimate numbers of individuals based on very small samples of our world—snapshots of the kind provided by moth traps. We have good estimates only for the very best-known groups of organisms. A few years back, my colleague Kevin Gaston and I tried to estimate how many individual birds there were in the world. Birds are undoubtedly the best-known major group of species—they are generally quite conspicuous animals, detected relatively easily by sight and/or sound, with a global network of keen (not to say fanatical) birders who aim to find as many as possible. There are many recording schemes, and numerous estimates of the abundance of birds at different scales, from the density of individuals in small patches of habitat to national population estimates. For example, the latest work suggests that the breeding population of British birds is 161,211,593 individuals— though this precision belies a substantial margin of error, and excludes nonbreeding individuals, the numbers of which are harder to assess. Pulling together data from a range of sources, Kevin and I estimated a global breeding population in the range of 100–400 billion birds, though we later revised this estimate down to a best guess of around 87 billion (which would have been around one 110 billion before humans began the process of converting natural habitats for our own use). This seems plausible; a recent study using different methods comes up with essentially the same answer, and the number is unlikely to be ten times smaller or ten times greater, at least. To "within an order of magnitude" (a multiple of ten) is often a reasonable approximation in ecology.

In the case of other organisms, it's much harder to be sure even of

---

still argue about the definition for eukaryotes). The number of prokaryote species may be in the rounding error of the number of eukaryote species, or perhaps the reverse is true. As viruses are parasites of prokaryotes and eukaryotes, the number of virus species is likely to depend on how many species of hosts there are, combined with how specific to any given host any given virus turns out to be (which will itself be variable). It seems barely worth attempting to calculate viral species richness, given the circumstances.

vi. This is the number reported at the time of writing.

the order of magnitude of estimates. According to the Smithsonian BugInfo website, the number of insects alive at any one time has been estimated to be around ten quintillion (one followed by nineteen zeroes). Where this estimate comes from, and whether it's reasonable or not, are hard to say. It suggests that there are more than a hundred million insects per breeding bird, which is perhaps plausible. For scale, both Great Tits and Blue Tits (European relatives of the chickadees) may deliver a caterpillar every minute to their broods at the height of the breeding season (an exhausting sixteen-hour day). Given that there are around 2.7 million pairs of these two species in Britain alone, that adds up to more than two billion caterpillars fed to nestlings of just these two species in just one day. Many of these will be the progeny of two species of moth—Winter Moth and Green Tortrix—which are key food items for the tits. Insectivory is a common diet for birds, and so supporting them and their hungry broods certainly requires a lot of insects. Insects are also a key food for many mammal species—a bat may catch 500 insects an hour—not to mention reptiles, amphibians, fish, spiders, and so on. Ten quintillion starts to seem ballpark. Even this number shrinks in comparison to estimates for microorganisms, of which a billion may be found in a teaspoon of soil. The estimated number of viruses worldwide is 1×10^{31}, or well over a billion for every one of those quintillion insects. Laid end to end, they would measure out a hundred million light years. Again, these estimates come with caveats over accuracy, but changing the numbers by even several orders of magnitude doesn't alter the message: this planet is home to a stunning abundance and diversity of life.

Of course, the second it's complete, any scan of our planet is out of date. New individuals will have been born, and others will have died. If those deaths involved the last of their kind, populations will have disappeared, and maybe species, too. Perhaps the births will have led to the gain of new species, although the nature of speciation is such that it is much harder to pinpoint the moment of appearance, versus loss. Regardless of births and deaths, individual organisms will have moved—huge numbers of them will have changed location. In absolute terms, these movements may not amount to much from second to second, but as time accrues, they will lead to areas being vacated or colonized by increments, and new species arising. A second later and the scene has changed again. This is a play that has been running since the first organisms appeared on earth. After almost four billion years, it has given us the planet we look out onto now.

None of these changes happens in isolation. Every individual organism needs resources to survive and reproduce—energy, water, and nutrients. Some will satisfy those needs directly from the environment, but for the majority, sustenance will come from other organisms, through consumption, depredation, or parasitism. These are the interactions that some consider a defining feature of ecology, and they mean that no organism plays out its time independent of others. They set species against each other, with the profits of some gained at the expense of others. Still others will have to work together for the mutual benefit of both. As a result, all of those billions of billions of organisms are locked in a dance, their mind-boggling numbers dwarfed by the numbers of potential and actual threads connecting them. The threads pull the organisms in myriad directions as they chase their needs across the environment, or try to evade the needs of others. All of these interactions happen against a backdrop of an environment that is itself constantly on the move, as geology and climate (and occasionally astrophysics) work together to alter the stage on which life plays.

It is not only the stage that changes—the species acting on it do, too. What even *is* a species? Many definitions have been proposed, and indeed there are several different philosophical approaches to the problem, but for practical purposes we usually consider a species as a group of organisms that can potentially (if of the right sexes) breed together in the wild to produce fertile offspring. The reason that defining species is hard is that they are dynamic. They respond to changes in the pressures imposed by their environment and from the other species with which they interact, with incremental changes of their own—that is, evolution. Species adapt and persist, but different pressures in different locations mean that some groups—populations—can head off on different trajectories, leading to splits, and, ultimately, to new species arising. How far populations have progressed down this road determines whether their members can still interbreed with other populations, muddying the waters about whether the different populations are really different species. The Deep Brown Dart and Northern Deep Brown Dart both breed in Britain, and as their names imply they are very similar. But are these moths different species? Even the experts cannot agree. Nevertheless, for the most part, it is clear to which species an organism belongs (with the caveat that we think most species are yet undescribed).

The process of gradual change and separation has, over some 3.7 billion years, given us the millions (or perhaps billions) of species that exist today. Each one has followed its own route across the ever-changing environment of Earth in an unbroken chain of descent from an ancestor common to all. The unique routes mean that each species is itself also unique in both its history and outcome: every species has devised a different solution to the challenges of surviving and reproducing, resulting in the characteristics they now display. Some species live at a rapid pace, growing quickly, reproducing as much and as often as they can, buying thousands of tickets in the lottery of life in the hope that some of the numbers will be winners. As we'll see later, if conditions are right, they can win big. Other species play a long game, taking their time to mature, putting a lot of time and effort into raising a few offspring to carry on the family line and potentially increase the species' representation in the global inventory of life. This is the approach we take as humans, showing that a long game can still lead to big wins. Many species run strategies somewhere in between. All individuals, however, make (unconscious) choices about how to allocate the resources they acquire to best ensure the persistence of the genes they carry through future generations. These choices are reflected in what the species look like, and how, in the broadest sense, they behave.

Individuals, populations, and species, their needs, interactions, movements, characteristics, and the ways they change—entities and events numbered in the quintillions—these are the purlieu of ecology. Ecology is sometimes criticized as a science for its lack of immutable laws and solid predictions, but the complexity of what it has to explain is truly mind-boggling. Charles Darwin used the analogy of "an entangled bank, clothed with many plants of many kinds, with birds singing on the bushes, with various insects flitting about, and with worms crawling through the damp earth . . . produced by laws acting around us,"[4] and for much of its lifetime, ecology has focused on trying to tease apart mechanisms at the scale he described. But an entangled bank is not separate from the wider environment in which it is set, and its plants and animals are affected by processes operating at the scales of continents and millennia. The moths that appeared in my trap that July morning, their numbers, identities, and characteristics, represent the denouement of one plot line in a play of daunting complication. How does one even begin to approach the question of how they ended up there?

The answer to this question at least is clear—by breaking the problem down into more manageable parts. All science works this way. The origins of physics, chemistry, and biology were responses to complexities too great to be tackled as a whole. We have continued to subdivide science as the expanding sphere of knowledge has stretched the boundaries of these traditional disciplines to the breaking point. Biology spawned ecology, but also evolutionary biology, genetics, cell biology, molecular biology, and other subfields too numerous to list. These are subdivided too, although our ultimate goal is to bring them back together into a holistic understanding of life.

At least for the moth trap, and for the science of ecology that helps us to understand the magic that it conjures, bringing some of the divisions together is what I will try in part to do here. Ecology is already a gigantic field, and while I've been studying it for around thirty years now, I make no claim to understand more than a tiny part. Nevertheless, some of that understanding does bear on questions of why, on any given day, my moth trap turns up the species it does, and in the numbers it does. The moth trap and its catch set me reflecting on these questions, and indeed on what we know about the natural world of which we are a part.

This book is the result of that reflection. I want to explain how we think the natural world works, or at least offer my take on those workings, through the window that the moth trap provides into a hidden world. There have been lots of books about particular animals, by scientists and by nature writers, and they are often wonderful. But *The Jewel Box* isn't a book about moths—or isn't *only* a book about moths. Rather, I want to use moths and my love of them as a tool to reveal the workings of nature. Just as Michael Faraday's iron filings arranged themselves to illustrate a magnetic field that would otherwise have been invisible, I want to show you that when we pay proper attention to these tiny animals, their relationships with one another, and their connections to the wider web of life, a larger truth about the world gradually emerges into focus. A single line of dialogue does not make sense without the rest of the play, and one cannot understand the contents of a moth trap without considering the complete environmental narrative. The contents of one small box depend on—and can illuminate—the workings of all of nature.

I start out simply, by thinking about a group of individuals belonging to just one species. The two fundamentals in the life of each one of these individuals—they are born, and then they die—contain within their interaction the capacity for populations to grow. This growth is the basis for understanding all of ecology, and indeed all of life's incredible diversity. But this growth happens on a finite planet, where resources are ultimately limited. This *really* matters.

Populations do not exist in a vacuum, of course, and no species plays out its lifespan independent of others. How these interactions modify the capacity for populations to grow informs the next two chapters. Competition between species for vital resources. And predation, when one species becomes the vital resource, consumed by a consumer. These processes help us to understand why populations do not grow out of control.

Birth and death bookend the life of every organism, but it is what they do in between that makes them the species they are. How they funnel those vital resources into growth, survival, and reproduction determines their life history—whether they will live fast and die young, or experience old age. The choices they make are driven by how and when death finds them, and lead to the diversity of forms the moth trap reveals. They help explain why there is no one right way to be a moth.

Having considered the processes that affect the ebb and flow of populations, and how those same processes help dictate life histories, I then step up a level of complexity to think about multiple populations living together, hence forming ecological communities. This starts to get at the key questions of how many species—and how many individuals of each—coexist. It's an open question whether interactions are king here, or whether communities are just random sets of species assembled by chance. I'll argue that the answer probably lies somewhere in the middle—members playing by rules, but membership depending also on a heavy pinch of luck.

Thinking about the structure of ecological communities highlights the importance of migration. My trap catches moths born in neighboring gardens and neighboring countries. Life depends on these movements. Migrants can colonize new areas and rescue dwindling populations from extinction. Much of the world would be a barren wasteland without them, and all of it less diverse.

The set of species that coexist in any given community is a subset of those in the wider environment. But the more species there are in that wider environment, the more species will tend to coexist. Not all parts of our planet are created equal when it comes to species, and I will discuss why. This brings us back to the inevitable bookends of birth and death, but now at the level of the species: speciation and extinction. Millions of years of this (with a bit of migration thrown in) have given us more moths in some regions than others, and more moths in some taxonomic groups than others. We cannot understand why a moth trap reveals the diversity it does without these widest of perspectives.

This is not quite the end of the story, as our tale has a final twist. A new actor has recently appeared on the scene, and is insinuating itself into every thread of our plot. That actor is an increasingly important driver of the processes that determine the workings of ecological systems, from the dynamics of single populations to the structure of communities and the diversity of whole regions. Its machinations threaten to send the whole story of life in a new and unwelcome direction, and I cannot finish without mention of its impacts—or to be precise, *our* impacts, for that new actor is us.

I hope this book will show that we cannot understand what goes on in our gardens, or on our roof terraces, in isolation. We can spend a lifetime describing in minute detail the environment, the organisms, and the interactions happening on an entangled bank, but this hard labor will be for nothing without context. All of nature is linked. The processes that determine the numbers and kinds of moths in my trap can depend on what my neighbors did last week, but can also span continents and eons. One cannot fence off a piece of nature and expect it to thrive, or even to survive. This realization has never mattered more. We are increasingly fashioning the world in our own image, remaking it into one where nature is limited to ever smaller pockets, set in landscapes dominated by humans and the processes we impose. If we care about our local nature, we need to think and act globally. The moth trap is a jewel box in which we can find Emeralds, Pearls, Rubies, and Gems. But the jewels are also pixels in a much larger picture. I hope I can convey something of what this picture looks like, and how it came to hold its current form. It is beautiful.

Chapter 1

The Gypsy Moth
The Power of Reproduction

All progress is based upon a universal, innate desire on the part of every organism to live beyond its income.
— Samuel Butler

Gypsy Moth, Camden, London.

The first night I ran the trap on my roof terrace in London, it turned up some exquisite moths. A Tree-lichen Beauty with a vivid turquoise stole draped across its shoulders. Jersey Tigers, their black-and-white-striped upperwings concealing rich orange underwings spotted with black. A self-explanatory Small Purple-and-Gold. Not everything in the trap that morning could reasonably be described as beautiful, though. One moth in particular had clearly seen better days.

Lepidoptera as a group are characterized by the microscopic scales that cover their wings (their name literally translates from the Greek as "scale wing"), which are the source of the incredible range of colors and patterns that moths display. The scales can easily be rubbed off, however. The wings of this moth had been scraped virtually bare, leaving blank canvas-like planes latticed with prominent, ridged veins. Lepidopteran scales can also be modified into hairs, and it is these that give many species the appearance of being wrapped in fur coats. This moth was virtually naked, its body hairs having been rubbed away to the extent that its thorax and abdomen were nearly smooth. Many moths are relatable through their resemblance to tiny flying birds or mammals. This one was a reminder that hidden beneath their fur is the hard chitinous exoskeleton possessed by all insects.

Stripped of most of its colors and pattern, the naked moth would have been difficult to put a name to had it not been for one highly distinctive feature that remained: a pair of broad, feathery antennae on its head, like tatty rabbit's ears. They were drooped and battered, but they pointed clearly to an identification. This was a Gypsy Moth. Scientific name *Lymantria dispar*, meaning "separate destroyer."[i] A male.

---

i. Throughout the book I call species by their "common" English names, but all also have a two-part scientific name. The first part denotes the genus (i.e., group of closely related species) to which the species belongs, while the second is unique to the species within the genus. The genus name is capitalized, the species name not, and both are italicized. Humans are no exception to these rules—science names us *Homo sapiens*. There are other species recognised within the genus *Homo*—our closest relatives, now sadly all extinct. A species can have many "common" names in a language (and there are moves to rename the Gypsy as the Spongy Moth because of the offense that "Gypsy" might cause to people of Romani origin), or none at all (such as *Enicospilus inflexus*,

The Gypsy Moth is a species that is naturally distributed widely across Europe and Asia, but one that has had a checkered history in Britain. Until the early years of the twentieth century, a small population clung on in a restricted area of the East Anglian Fens, where its caterpillars fed on a couple of shrubby plant species, Bog Myrtle and Creeping Willow. Unfortunately for the Gypsy, the Fens have deep and fertile soil, and have long been coveted as farmland. Today, almost all of the original marshy habitat there has been drained and put to growing food, including the areas where the last surviving English Gypsy Moths lived. Now, satellite images show a grid of fields, the straight edges so typical of human influence on the environment. The Gypsy Moth was last seen in its Fenland refuge in 1907.

As it turns out, this was not the end of the story for the Gypsy Moth in England. In the summer of 1995, the species was discovered in Epping Forest, one of London's green lungs, to the northeast of the city. This was not an outpost of the original population being unearthed, but a new population based on colonists from the continent. We know this in part because, while the Fenland Gypsys were rather picky eaters, the Epping Forest caterpillars are much more catholic in their diets—technically, *polyphagous*. Exactly where the colonists came from, and when they first arrived, is unknown, but they almost certainly hitchhiked into the country on imported wood products such as timber or packaging material. The female Gypsy Moth is largely flightless, and so tends not to move far once it has eclosed from its pupa. They lay eggs in large yellow clumps (called *plaques*), normally on trees, but in modern times also on fences, walls, and other solid surfaces. Eggs laid on trees or wood subsequently cut for timber or pallets could easily ride the ferry or Eurostar from the continent to hatch out across the Channel in southeast England. The fact that the new population is mainly found in economically active London is consistent with the idea that the moths arrived inadvertently on cargo, rather than being natural colonists. Either way, since that first sighting the population

---

which we will meet later), but it only ever has one valid scientific name. Scientific names can get changed, though (the Gypsy Moth started out as *Phalaena dispar*, for example), which is a pain.

has grown and spread. It is now found throughout the London area, and beyond. More than a century after the last individuals disappeared from the East Anglian Fens, Gypsy Moths are back again in England, and (the males at least) appearing on my roof terrace in Camden.

While the precise origin of the Gypsy Moth population in London is uncertain, that isn't the case for the population in the United States. They came from 27 Myrtle Street, Medford, Massachusetts—the house of Mr. Léopold Trouvelot—in either 1868 or 1869.

Gypsy Moths do not occur naturally in North America.[ii] They were taken there by Léopold after a trip to Europe as part of his experiments on the production of moth silk, a valuable commodity then and now. Léopold was himself a native of Europe, described in the 1890s as an artist, naturalist, and astronomer of note, but now mainly remembered for his role in the Gypsy Moth saga. He probably imported the moths as eggs, and then either some of those eggs or the caterpillars that hatched from them were accidentally blown out of the window of the room where they were being kept. Realizing that the consequences of this could potentially be severe, and unable to put the worms back into the can himself, Léopold apparently gave public notice of the escape. Exactly what is meant by that is unclear. In the circumstances, it doesn't really matter.

The genus to which the Gypsy Moth is assigned has changed over the last 150 years, but a common theme of its name in translation is destroyer or ravager. The polyphagous form can feed on a wide variety of tree and shrub species, and can cause significant damage to individual plants if population numbers get high. Hence Léopold Trouvelot's apparent consternation at having some of his stock escape. However, for the next decade it seemed as though it would turn out to be an accident of little consequence. Léopold evidently saw Gypsy Moths,

---

ii. Species that are moved by humans to places beyond the normal limits of their distributions, and that are released or escape into the wild in those places, are termed *aliens*. The Gypsy Moth in Medford is a classic example, but we will hear much more about them later.

presumably in or around his garden, but hardly anyone else did. That started to change as the moth entered its second decade in Medford.

In 1879, a gentleman by the name of William Taylor moved into 27 Myrtle Street, Trouvelot having moved on by that point. The following spring, Taylor "found the shed in the rear of his house swarming with caterpillars." They were such a nuisance that he got permission to sell the shed, thereby presumably moving some of the caterpillars to a new location.

Within a couple of years, neighbors of no. 27 were starting to feel the effects of the moth, too. There were caterpillars all over the outside of no. 29, and their apple and pear trees were entirely stripped of leaves. Gypsy Moths continued to spread along Myrtle Street and into natural areas to the south, but it wasn't until 1889 that the full extent of their capabilities became apparent. The outbreak of Gypsy Moths in Medford that year was so devastating that the locals wondered how on earth the species could have gone largely unnoticed for almost two decades.

In our times of diminished nature, it may be hard for many of us to conceive of the abundance of the Gypsy Moth in the outbreak of 1889. Yet testimony to the fact is itself abundant, as documented in a report on the problem to the Commonwealth of Massachusetts published in 1896. It tells how some trees were so covered with the egg plaques that they appeared spongy, and yellow in color. The eggs produced caterpillars in millions. One Mrs. Belcher reported that "My sister cried out one day, 'They [the caterpillars] are marching up the street.' I went to the front door, and sure enough, the street was black with them, coming across from my neighbor's, Mrs. Clifford's, and heading straight for our yard. They had stripped her trees."

The caterpillars were nicknamed "army-worms," and for good reason. Mrs. S. J. Follansbee related that "the walks, trees, and fences in my yard and the sides of the house were covered with caterpillars. I used to sweep them off with a broom and burn them with kerosene, and in half an hour they would be just as bad as ever. There were literally pecks of them.[iii] There was not a leaf on my trees." Mrs. Snowdon: "I have seen the end of Mrs. Spinney's house so black with caterpillars that you

iii. One peck is around nine liters.

could hardly have told what color the paint was." Mrs. Ransom: "In the evening we could hear the caterpillars eating in the trees. It sounded like the clipping of scissors." Mr. Daly had a similar experience: "At nighttime we could hear the caterpillars eating in the trees and their excrement dropping to the ground." Walking around town at this time was distinctly unpleasant, while hanging out washing to dry simply risked having to wash it again—it would often come off the line dirty with frass. The army-worms did not stop at trees, either: "When the supply of leaves in the trees fell short (and oftentimes before) they attacked the gardens. Little was spared but the horse-chestnut trees and the grass in the fields, though even these were eaten to some extent." Dealing with the problem was a significant undertaking, as noted by Mrs. Hamlin: "For six weeks a great deal of our time was devoted to killing these caterpillars."[1]

If the authorities had not been alert to the presence of the Gypsy Moth before, they certainly were now. Before 1889 was out, state politicians were being urged to act against a species that was clearly a threat to both forestry and agriculture, not only in Massachusetts but countrywide. The state government moved quickly, and by March 1890 the first funds for the eradication of the Gypsy Moth had been approved. Unfortunately, assessments that year found that around fifty square miles were already infested. Control within that area was quite successful in reducing Gypsy Moth numbers through manual destruction of eggs and spraying of affected trees with Paris green, a highly toxic compound of copper and arsenic. Nevertheless, the infested area continued to grow. While initial extermination events had focused on trees, masses of eggs had also been laid on fences, under boardwalks, under steps and in cellars. Searching these and other areas turned up more than 750,000 plaques in just the first six weeks of 1891, or on the order of 300–500 million eggs.

The army-worms were well and truly out of the can. Starting from Léopold Trouvelot's garden, the Gypsy Moth has now spread to occupy more than 400,000 square miles of northeastern North America. Its population size varies, and in some years numbers are low. In the outbreak year of 1990, though, around 100,000 square miles of American forest was defoliated by its voracious caterpillars. Ravager indeed.

One rather tatty moth in my first night of trapping, but so many questions. What was the Gypsy Moth doing in Medford for the decade when it passed largely unnoticed? What happened after two decades to cause the population there to explode? Could that happen in London? Why has it done so well in North America—a part of the world that it didn't evolve to inhabit? And why does its North American population seem to have good years and bad—why are there years when its abundance reaches plague proportions?

At first sight, these are daunting questions. Consider all the factors that could be affecting Gypsy Moth populations. There are all the vagaries of the environment. Temperature and rainfall—their highs and lows, variation across the day and year, and the effects of unexpected extremes. Geology (which affects soil) and edaphology (how soils interact with living organisms) vary, too—does that even matter for moths? Then there are all the other organisms with which the moths are sharing the environment—species that they may consume or that may consume them, competitors for resources, species that help or hinder their growth and reproduction. Features of the moths themselves may matter, and can change through evolution or phenotypic plasticity (changes in response to an environment that do not require evolutionary change).[iv] All of these possibilities and more may drive Gypsy Moth numbers up and down. How do we pick this all apart?

Some of these questions are the subject for later chapters, and I will try to answer them then. However, before we can begin to come up with those answers, we need to understand how populations change in size—how populations grow, and how they decline. We also need to understand what we mean by a "population." Fortunately, the answers to these questions are relatively simple. The consequences of the answers can be less so, but let's start with the simple bits.

For an ecologist, a population is simply a set of individuals, all of

---

iv. For example, caterpillars of the American Emerald Moth *Nemoria arizonaria* mimic the oak catkins they feed on in the spring, but twigs when they feed on oak leaves in the summer. Their diet determines which sort of mimic they grow up to be, and the different caterpillar morphs were originally thought to be different species.

the same species, living in a defined area, at a given point in time. I say "simply," but that definition already presents difficulties.

In some cases, it can be surprisingly hard to distinguish one individual from another—two flowering stems might be shoots from the same root stock, or they might be neighbors. This can make counting the individuals difficult when we want to measure the number of individuals in the population—*population size* (which we will). For moths, at least, individuals are quite easy to define, at any stage in their life cycle. But they can still be hard to count.

The area delineating a population can be more difficult to define, as the commission established by the Commonwealth of Massachusetts to deal with their Gypsy Moth problem quickly found. Generally, we can sidestep the issue of precise boundaries and define it arbitrarily, for the convenience of the researcher—the area is the area that we're interested in, which, in general, it is (though this does have consequences, as we will see in a moment). We do need to be careful about our choice, however. Too small an area, and the numbers of individuals there will be too small to say much about the population and its workings. Too large, and there will be too much information to process—counting all the individuals, or even estimating the number, becomes a headache.

How we think about a "point in time" is largely determined by the population we are studying. Time moves differently for people and moths, while ecologists following bacterial populations in petri dishes will work on different timescales than their colleagues monitoring elephants in the national park down the road. What matters here is that we follow populations over periods that allow us to understand what might be determining population size, and for that, we need to pick time steps that allow us to track changes in that size.

Once we have defined our population, we can then try to understand if and why the number of individuals comprising it changes. This is the study of *population dynamics*. For this, we first need to count or estimate how many individuals there are in the population. We then need to do this many times, so that we can follow change. We don't need to census the elephant population every day for this. We probably *do* need to for the bacteria. Studying bacterial populations in petri dishes will obviously give us data to explore population dynamics more quickly. Whatever our population, though, we need to understand how it changes in size. And there is a huge diversity of processes that can affect this—environmental, ecological, and evolutionary—making it a daunting proposition at first glance.

Luckily, despite all the different processes that can affect population size, ultimately it changes through the action of just four basic processes: birth, death, immigration, and emigration.

Take the Gypsy Moth population in Léopold Trouvelot's garden as an example. It could get larger because more Gypsy Moths are being born (eggs being laid, caterpillars hatching, and adults eclosing, depending on which life stage you're counting). It could also get larger because Gypsy Moths from elsewhere are moving into the garden—this was how the population started, in fact, through immigration from the house. However, the population could also get smaller because moths are dying—this is the fate of all members of all populations eventually, and was soon a major objective in the garden of 27 Myrtle Street and its environs. Finally, the Gypsy Moth population in Léopold's garden could decrease because moths are moving out. Such emigration certainly happened in the 1880s, and likely before that, too—and how! One garden's emigrants are another's immigrants.

Year on year (as an appropriate time scale), then, the Gypsy Moth population in Léopold Trouvelot's garden could have increased through births and immigration, and decreased through deaths and emigration. I'm now going to write this information in the form of an equation. I am no mathematician, and the equations that follow will involve no more than addition, subtraction, and multiplication (there will be one Greek letter). So:

If we assume that 1869 was "Year zero" for this population, then by the following year:

$$N_{1870} = N_{1869} + B - D + I - E$$

In this equation, N is the number of individuals in the population (at the end of 1869 and the end of 1870, as denoted by the subscripts); B is the number of individuals born into the population present in 1869; D is the number of individuals in the 1869 population (or added to it in

1870) that then died before 1870 was out; I is the number of individuals that immigrated into the 1869 population from elsewhere; and E is the number of individuals that emigrated out of the 1869 population for pastures new. The change in the population size between 1869 and 1870 depends straightforwardly on the numbers of births and deaths, immigrants and emigrants. We can generalize the equation above by replacing the specific years for general time steps (t and t + 1):

$$N_{t+1} = N_t + B - D + I - E$$

This equation is quite a simple one, though not as simple as it could be. We are actually distinguishing different types of process going on in it—two that are fundamental to all life (birth and death, B and D) and two that are not, but that relate to movement *between* different populations (immigration and emigration, I and E). If we assume that our population is *closed*—that is, individuals neither enter nor leave it—then we can dispense with the movement part. In that case:

$$N_{t+1} = N_t + B - D$$

This equation would apply to the size of the Massachusetts Gypsy Moth population in 1870—only births and deaths matter. (With the caveats that we are assuming that no more individuals escaped through Léopold's windows, and that we would soon be having to expand our area of interest beyond Massachusetts to continue to ignore emigration.)

The equation for the change in the size of the population between time steps (i.e., the difference in size between N_t and N_{t+1}) is simpler still:

$$\Delta N = B - D$$

Here, the triangle is the Greek letter delta, the symbol ecologists use to mean "the change in." So the change in the size of a population between time steps—Gypsy Moths in Massachusetts between 1869 and 1870, say—is simply the number of births minus the number of deaths. If there are more Gypsy Moths born in Massachusetts in 1870 than die, then the population will increase (ΔN will be a positive number). If more die than are born, then it will decrease (ΔN will be a negative number).

I am going to persist with the algebra through a couple more steps, because they are baby steps, and because the outcome is fundamental in biology. I also think the result will be of interest, at least at the time of writing, so please bear with me a little longer.

B and D are the *number* of births and deaths in a population. What is going to influence the magnitude of those numbers for any given population? The most obvious factor is how many individuals there are in the population. Larger populations (bigger N) will generally have more births and more deaths, all else being equal. More people are born and die each year in the United States than in the United Kingdom, just because the population of the United States is much larger. The same is (now) true of the Gypsy Moth populations in these two countries.

What we can do, then, is rewrite B in terms of the size of the population (N) and the average number of offspring per individual in that population, or the birth rate, which we denote by lowercase b. Thus, $B = bN$ (which means b multiplied by N—we don't write the "times" sign to avoid confusion with the letter x). We can do the same for the number of deaths, D, which is then the product of the size of the population and the average number of deaths per individual in that population (or, less confusingly, the proportion of individuals in that population that die—the death rate), d: $D = dN$. This gives us:

$$\Delta N = bN - dN$$

The change in the size of a population between time steps (ΔN) thus depends on the size of the population (N), and the difference between the birth rate (b) and the death rate (d) in that population. If an individual is more likely to die than to produce an offspring, the population will shrink. The fact that $bN - dN$ is mathematically the same as $(b - d)N$ emphasizes the point: b has to be greater than d for the population to grow. For the Gypsy Moths of Massachusetts, this was very much the case.

The final edit we make to our equation is to give $(b - d)$ its own designation, r:

$$\Delta N = rN$$

If r is a positive number (the birth rate is higher than the death rate), the population will grow. If r is negative (the death rate is higher than the birth rate), it will shrink (ΔN is a negative number, so we are subtracting individuals from the population). An r of zero, and the population stays constant. We call r the *intrinsic rate of increase* of the population. It's sometimes called the *Malthusian parameter*, after the eighteenth-century cleric and scholar Thomas Malthus, who famously wrote on questions of population growth, and whose work influenced the evolutionary ideas of Charles Darwin.

And that is the end of the equations for this book.[v]

Perhaps the concept of a population "r" number rings a bell? In the spring of 2020, this letter was in the news in a way that I could not have imagined when I had lectured to students on this theory of population change a few months earlier. The capital R that is so important to the trajectory of the Covid-19 pandemic is not quite the same quantity as the Malthusian parameter (for R, 1 is the critical value, not 0), but it is the equivalent for the population of viral hosts (determining changes in how many of us humans are infected). And the outcome of a constant value of r above 0, or R above 1, is the same in each case—*exponential* population growth. This is a concept that strikes fear into the heart of epidemiologists, or pest managers, because it essentially means unconstrained population growth. Growth out of control. The consequences of this process can be staggering.

The power of exponential growth is illustrated by the well-known legend of the Indian king and chess enthusiast challenged to a game by a sage, who was actually the god Shiva in disguise. The price of the inevitable defeat for the king was some grains of rice—he had to place one grain on the first square of the chessboard, two grains on the next, four grains on the third square, and so on, doubling the number of grains on each square up to the total of sixty-four on the 8 × 8 board. The small numbers on the first few squares belie the power of doubling in this way. The square at the end of the third row already requires more than eight million grains of rice from the king. The sixty-fourth square

---

v. Except for footnotes! Feel free to skip them.

needs more than nine quintillion (that's 9,000,000,000,000,000,000, or 9×10^{18}). The board as a whole requires somewhere in the range of 400–500 billion *tonnes* of rice, many times the annual global harvest.

The scientist and blogger David Colquhoun prefers the analogy of a leaking pipe in Wembley Stadium. The pipe leaks one drop of water as the cup final kicks off, two drops after 1 minute, four drops after two minutes, and so on. The entire stadium overflows before halftime.[2]

In terms of our last equation above, the population of rice grains is changing from square to square (ΔN) with $r = 1$: the next square gets an extra rN (i.e., N) grains, where N is the number of grains on the previous square ($1 + 1 = 2$; $2 + 2 = 4$; etc).[vi] For a population of living organisms, this is equivalent to a situation where, before it dies, each individual in the population on average gives birth to two offspring that themselves survive to reproduce. That doesn't seem to be all that many.

The fecundity of a female Gypsy Moth depends on its size, but an adult can typically lay hundreds of eggs. The scaling from number of plaques to number of eggs made by the scientists studying the early outbreaks in Medford works out at an average of around 400 eggs per plaque. Not all of the eggs will produce caterpillars, not all of those caterpillars will survive to pupate, and not all of the pupae will hatch into healthy adult moths. Not all adults will themselves reproduce. Yet, if only 1 percent of those eggs becomes a reproductive adult, that implies an r of 1 (assuming that half of the offspring are males, which don't lay eggs). Clearly, r for the Medford Gypsy Moths in those early days of its population growth could have been much greater than 1, and the population more than doubling each generation. Even with r less than 1, though, population growth would have been inevitable (as long as r was greater than 0), just slower.

---

vi. In fact, this is not strictly correct. On the chessboard, the individual squares are *discrete* entities, and the increases in numbers of rice grains happen in a series of separate steps from square to square. Our equation for exponential growth is *continuous*—even though we think about total changes from year to year, say, population growth can happen every day (or hour, minute, second). This means that r is not exactly 1 for the chessboard analogy. The precise numbers matter less than the general concept, though, so we can let it slide. There is in fact an equivalent equation for discrete exponential growth: $N_{t+1} = \lambda N_t$. In the example here, $\lambda = 2$.

Exponential growth is our basic model of how a population grows (or declines) in ecology. It's simple, but its implications are clear. It describes why a population grows, and why that growth can lead to populations of plague proportions. It may also explain why population growth can appear to creep up on us—for example, why it took a decade or so before Gypsy Moths came to be noticed outside of Léopold Trouvelot's garden at 27 Myrtle Street. At first, a population growing exponentially can appear to be growing quite slowly, especially if r is not very high. Think about the king's chessboard—the last square on the first row needs only 128 grains of rice to be deposited. That is seven steps from square one, which you could equate to seven generations of a biological population with an r of 1. If the rice were moths, a population of just 128 would barely register. This feature of exponential growth can explain why populations like the Gypsy Moth in Massachusetts are easily overlooked at first—this "lag phase," when numbers remain quite low. A further seven generations at the same growth rate, though, and the population would be just over 16,000. This many moths would certainly get noticed. Seven generations more and the number is more than two million. Trees in the neighborhood would already be taking a pounding.

We rarely get to see unconstrained growth in biological populations, and the early years of the Gypsy Moth in America provide us with as good an illustration of the phenomenon as we could hope for. (Not that it was appreciated by the good folk of Medford.) Another prime example followed the release of twenty-four European Rabbits in Victoria, Australia, in 1859 for hunting. By the 1920s, the Australian rabbit population was estimated to be of the order of ten billion animals. This is why epidemiologists were getting worried early in 2020 when Covid-19 case numbers still looked low—they knew the power of exponential growth. The direction of travel of the pandemic mattered more than the numbers infected at that point. Those numbers were growing exponentially. The capacity for this sort of increase is intrinsic to all populations. The more important question is: Why do we see it so rarely?

The answer to this question is also simple—we rarely see unconstrained growth because, eventually, growth does become constrained.

The exponential model assumes that the value of r never changes: however large the population gets, it continues to grow at an ever-increasing rate (because change, our ΔN, is determined by r and N, and N gets ever larger). In the legend, the unfortunate king has to keep adding more and more rice to squares on the chessboard, until the last square, which needs more rice than the annual worldwide harvest. And this of course is where the exponential model falls down—eventually growth becomes limited by some other factor than the availability of individuals to reproduce. We live on a finite planet. The king runs out of rice. Caterpillars run out of leaves. Population growth hits the buffers.

What this means in practice is that as a population grows—the Gypsy Moths in America, say, or the Gypsy Moths in their nineteenth-century Fenland refuges—r changes. This happens because one or both of its component parts changes with population size. Either the birth rate of the population goes down, or the death rate goes up. We can illustrate the effect straightforwardly. Let's say that $d = 1$, so every year all the adults die, but that each individual has on average two offspring ($b = 2$). Then, r is $b - d$, or 1, and the population grows. As the population grows, though, food becomes harder to find, and adults can't reproduce as much as before. The birth rate falls. Eventually, food becomes scarce enough that each individual can only find enough spare food to fuel the production of one offspring per year on average. Now b and d both equal 1, and so r has fallen to 0. The population stops growing. (It's easy to imagine conditions where this happens because the death rate increases instead.)

If we were to plot a graph showing how population size changes over time, exponential growth would look not unlike an italic J, with the upstroke getting ever steeper. A graph of the distance that Gypsy Moths were recorded from Léopold Trouvelot's garden (as a proxy for population size) versus time shows exactly this shape, albeit with a long lag phase. But if r decreases as the population grows, then eventually population growth stops, and the population levels out. On a graph, this growth curve would look not unlike an elongated S. We call population growth of this sort *logistic*. It flattens out when the population reaches the point at which the birth rate and the death rate are equal. We call this point the *carrying capacity* of the environment for that species. It is the size of the population that can be maintained given the finite resources available.

In ecology, we describe logistic growth as *density-dependent*. As the population gets larger, its density, which is the number of individuals divided by area, also increases (remember: we define a population on the basis of a specified area of interest). But as density increases, so *r* decreases—the value of *r depends* on density. This contrasts with exponential growth, where *r* is independent of density. The contrast between density-dependent and density-independent processes is a common theme in ecology. Density-dependent processes are typically how the brakes are put on population growth, because the pressure they exert increases as the population does, too.

Another way of thinking about logistic growth is as modified exponential growth. Initially, the two forms of growth are more or less identical, because small population sizes (or low densities) have little impact on *r*. It is only once the population gets large enough to have a noticeable impact on birth or death rates that growth starts to deviate from its ever-increasing exponential form, beginning to slow, and eventually flattening off.

The Gypsy Moth exploded into the consciousness of the residents of Medford in the 1880s, beginning its march across Massachusetts and eventually most of northeastern North America. Yet New Englanders have never been knee-deep in Gypsy Moths—although they may have been at least toe-deep in Medford in those early years. In most years now, Gypsy Moth populations are stable and low. Clearly, something has modified their exponential growth, and it is certainly plausible that their populations should have some sort of density-dependent control. There is indeed evidence of this, from observations and experiments in the field. Years or sites with higher densities of Gypsy Moths also tend to see lower birth rates and higher death rates in those populations. Females tend to lay fewer eggs when caterpillar densities are higher. Attrition rates are higher in the caterpillars, fewer of which survive to pupate.

The key cause of density-dependent die-off in Gypsy Moth caterpillars is disease—they get infected by nuclear polyhedrosis virus, or NPV. This is a species of baculovirus, a group primarily known to cause diseases in insects, and especially moths. It's thought that some caterpillars catch it young by eating the shells of the eggs they hatch from, which are coated with virus particles. Others catch it by eating contaminated vegetation. The virus moves into gut cells, and then into the cell nucleus. Here it reproduces, with the new virus particles spreading from cell to cell. Eventually the cells rupture, and the infected caterpillar dies as a bag of virus-packed fluid. The dead become a source of particles to infect others. Infectious diseases make hay in large and densely packed populations, where they can easily spread from host to host. As long as each host infects more than one additional host, the disease will spread, potentially exponentially. (Unless you spent 2020 and 2021 living completely off the grid, you will probably know this.) NPV is not the only killer of Gypsy Moths, but I'll talk more about predatory and pathogenic interactions between different species in a later chapter. Suffice to say here that it's not uncommon for a population of one species to encounter difficulties when it butts up against a population of another.

Whether the Gypsy Moths of North America have any lessons for their relatives in London is unclear. At present, the London population is slowly growing and spreading, but it is not a notifiable pest: we are not (yet) scraping them off our buildings with brooms and burning them by the peck. Presumably, something is keeping this population largely in check. In the real world, no population exhibits pure exponential growth. It will always be checked by some force. That said, nor does any population trace a smooth logistic "S" to a flat carrying capacity defined by density dependence in growth rates. The real world is far more complicated than these simple models portray.

Nevertheless, it's a well-known maxim in statistics that while all models are wrong, some are useful. I've been telling undergraduates this for years now, but the Gypsy Moth in my trap that first night prompted me to think more about these processes than I have since I first lectured on the topic. That tatty moth and the history of its relatives made me realize just how useful those models are. They underpin ecology. I have spent so much time on them—and even risked alienating you with equations—for that reason. Much of the rest of ecology consists of attempts to understand how species try to game the numbers between the inevitable births and deaths of their constituent individuals in order to do the best job of upping their abundance or their distribution. As we saw in the introduction, this is the very definition of the field.

This game also underpins evolution. Charles Darwin realized that it is the enormous capacity for life to reproduce that provides the pieces with which natural selection plays. Species will exploit their environment by turning usable resources into as many offspring as they can. Yet resources are finite, and inevitably times of feast will turn into times of famine. Some individuals will survive, but many will not. Any characteristic that allows an individual to leave more offspring than its fellow strugglers will be favored, and spread through the population. This process has the power to transform populations, but also to cause them to diverge in their traits when conditions differ. New species form. The interplay of birth and death rates gives populations the capability to grow out of control, but it is also the driving force for life's diversity.

The Gypsy Moths that escaped through Léopold Trouvelot's window got lucky. They reached a New World full of opportunities, and seized them. Just how lucky they were, though, needs an appreciation of just how precarious life can be.

So far, we have been thinking about population growth as *deterministic*. What that means is that the destiny of a population depends entirely on its r number. If r is greater than zero, a population grows—indefinitely in exponential growth, and up to the point where r dwindles to zero for logistic growth. Once r is zero, the population stays steady. If you know N and r, then you can calculate the change in population size over the next time period with certainty. The number of grains of rice on the next square of the chessboard is a given. But the real world does not behave like that. In reality, there is a major role for chance. A population that, on average, leaves more offspring in a given time period than it suffers deaths (i.e., its r is greater than 0) can still die out if a random event impacts either of those numbers.

Randomness (which, technically, we refer to as *stochasticity*) can take a variety of forms, but at this point, I'll focus on randomness in the environment—environmental stochasticity. We all know how variable the environment can be. I'm a Brit, so the weather is my main conversational gambit, and where I live we can experience its vagaries on a day-to-day basis. Yesterday was a lovely November day for a walk on Hampstead Heath, today very much one for coat and umbrella. Tomorrow has sleet in the forecast. The local flora and fauna are well adapted to this typical variation, which is one of the driving forces underpinning their distributions. However, it isn't the normal fluctuations that matter, but the extremes. A bad storm, a cold snap, a prolonged drought—these can easily interfere with breeding seasons, or increase mortality, to the extent that populations disappear. On average, a population might be able to grow happily in a certain environment, but not across the full spread of conditions there. (As an aside, this is why 2°C [3.6°F] of climate warming might not sound like much, but it greatly worries scientists. That's the average warming—the extremes will get much worse, everywhere. Heatwaves kill.)

Environmental stochasticity is a particular issue for small populations. When a population is starting out in a new area—when its numbers are low, and it hasn't yet spread far—it is much more susceptible to extreme events. A few extra deaths (a small increase in d) would make little difference to the current Gypsy Moth population of North America, but back in 1868 or 1869, when there were only a few moths in Léopold Trouvelot's garden, a bit of bad weather could easily have done for them. We know from similar introductions of birds around the world that an unusually large storm in the years immediately following their escape or release significantly increases the likelihood that the population will fail. A cold snap or a bad winter, or just a bit of extra effort at insect control by Monsieur Trouvelot, and this chapter could have been very different.

Environmental stochasticity is easy to picture in terms of the weather, but other unusual events can also eradicate populations. Random losses of key patches of habitat will do the job, for example through fire or flood. The precarious toehold that the Gypsy Moths had in Medford in those early years is illustrated by their long-lost family in the English Fens. The fussier eating habits of the moths in this population apparently limited them to just a few small patches of habitat that had survived the draining of the rest of the Fens, starting in the seventeenth century and accelerating through the eighteenth and nineteenth. Whatever the typical birth and death rates of the moths in this population, they could never survive the massive increase in death rate imposed by the loss of their habitat. This sort of impact is very much *not* dependent on population density.

Usually, random periods of bad weather, or accidental losses of patches of habitat, do not result in extinction. Populations tend to be large enough to absorb the hit and rebound. (Or, as we will see later, there are enough interconnected populations for immigration to come into play.) However, stochasticity is the reason why conservation biologists worry about very small populations of threatened species. A bit of environmental bad luck can easily drive a small population to extinction, no matter how carefully we try to look after it. And for a threatened species, that might be its *only* population. The last Fenland Gypsy Moths finally disappeared as the twentieth century dawned.

Natural fluctuations in the environment are a fact of life, and they cause fluctuations in populations of organisms in response. However, we can also get fluctuations without *any* element of stochasticity—when population growth is *entirely* deterministic. Some classic examples of this arise from the basic model of logistic growth.

All models come with a set of assumptions underpinning them— essentially, a set of more or less credible beliefs that we accept as true for the purposes of the maths. For example, the exponential and logistic models assume that populations are closed (there is no migration), which may often be true. They also assume that all individuals are identical, parthenogenetic (capable of virgin birth), and able to start reproducing as soon as they're born, which obviously is not true for most real organisms. I've glossed over these less credible assumptions because they don't affect the key insights from the models: populations have the power to grow out of control, but control eventually comes to them.

Logistic growth happens when adding individuals to a population causes r to decrease, either through a decrease in birth rate or an increase in death rate. In theory, eventually r reaches zero, and the population stops growing: it stabilizes at its carrying capacity. An additional assumption of the logistic model, though, is that the response of r to each new individual added to the population is instant. Each extra organism slows the birth rate and/or increases the death rate as soon as it comes into existence. This seems unlikely to be true, perhaps for any real population, but certainly for most. More plausible is that a response in terms of births or deaths would be delayed to some extent. Eventually the extra population would feed back into these vital rates, through a shortage of food, for example. The population growth rate would slow, but not instantly. There would in fact be a time lag in its response.

Time lags are built into the dynamics of many populations because of the seasons, especially in higher-latitude ecosystems where the year alternates between periods of plenty and periods of hardship. Population growth is not continuous over time. Individuals are added to the population all at once (more or less) in the spring and summer breeding season, but can die at any time. The annual flush of births means that r cannot react instantly to population size—the population size in any given year depends on what happened the previous year. The Gypsy Moth is an example of a species that lives this way.

Time lags have some interesting consequences for the logistic growth model. A delayed decrease in r following population growth can lead to the population overshooting its carrying capacity, ending up with more individuals than the environment can support. In this case, the population must decrease to carrying capacity—and the time lag can now lead to it decreasing too far. We start to get fluctuations in population size around the carrying capacity.

Exactly what these fluctuations look like depends on a combination of how long the time lag is, and how large is r. If the combined effects are small, then the fluctuations eventually dampen out. The population does not grow too fast or far beyond its limit, higher or lower, and soon settles down at its carrying capacity. If the combined effects are larger, though, the population can settle down into regular cycles. It overshoots carrying capacity far enough that it is then forced to undershoot when it declines. Overshoot follows undershoot again and again, *ad infinitum*. Good and bad years for the Gypsy Moth could, in theory, be driven entirely by the species' own behavior.

A population may establish regular cycles around carrying capacity, undershooting and overshooting by the same amount every time, but under certain conditions, those cycles can shift into chaos. The population continues to overshoot and undershoot carrying capacity, but to a different amount each time. The pattern of fluctuations doesn't repeat, and the size of the population each year looks completely unpredictable.

This unpredictability is illusory. The chaos here is entirely deterministic—if the starting conditions (r, N, carrying capacity) are the same, the population will always follow the same, apparently random, pattern of fluctuations. Yet a slight difference in the parameters—r is 2.91 instead of 2.9, say—and the population fluctuations follow a slightly different pattern. They would start out almost the same, but over time, any tiny differences in population parameters would eventually lead to large differences in population sizes. Even if the real world perfectly followed our logistic models, we would still not be able to predict the size of a population in, say, twenty years' time if we could not also calculate r or N perfectly. There is no margin for error.

As I mentioned earlier, the North American Gypsy Moth population is not stable, even allowing for the fact that it continues to spread across the continent. The population naturally varies from year to year, typically at low levels. Sometimes Gypsy Moths can be hard to find in the forests. This has some of the appearances of fluctuations around a carrying capacity, as expected from logistic growth. The reality, of course, is more complicated. Occasionally, the Gypsy Moth recapitulates its early history in Medford, and explodes to plague levels across large areas of its new range. Massachusetts saw around a million hectares of forest (almost 2.5 million acres) defoliated by this insect in 1981, with Maine, Vermont, and New Hampshire similarly hit. These outbreaks are unpredictable in their size and extent. So, chaos?

Probably not. We certainly are poor at predicting variation in ecological populations, but this is hardly surprising. Deterministic chaos shows us that even in predictable systems, small differences in where we start off lead to big differences in where we end up. We couldn't measure the necessary features of populations well enough to predict their fluctuations even given complete determinism. It is a salutary lesson that we would not be able to predict population dynamics even if they were completely predictable! Of course, the environment is inherently stochastic, and this random element just adds to our problems. It is testament to the determination of ecologists that they still work to identify order in the chaos.[vii]

A moth trap is a great way to lure out representatives of some of the populations that live lurking and unseen in the trees and undergrowth around you. Moths are adept at going unnoticed, until attracted by a light. I had no idea that there were Gypsy Moths where I lived until one materialized in my trap. People have been using lights to attract and catch moths at least since Augustus was emperor of Rome, but the kind of light-plus-box trap that I run on my roof terrace has only been around since the early twentieth century. I do wonder what Léopold Trouvelet would have caught had he been able to run one in his Medford garden in the 1870s, or indeed whether he had Gypsy Moths fluttering in through his windows on warm summer nights. Would his catches have tracked an increasing population? If so, perhaps he would have been more forceful in bringing their escape to public notice. Hindsight is, of course, a wonderful thing. Ecologists monitor populations—including through networks of moth traps—among other reasons, for just this sort of early warning. Of populations increasing or decreasing in ways that might flag a need for us to worry.

Populations grow when births outnumber deaths. The difference doesn't have to be great for the population to grow, although the longer it spends at low numbers, the more at risk it is at from the vagaries of

---

vii. In fact, they have found some evidence that the Gypsy Moth outbreaks follow a ten- or eleven-year cycle. I will return to the question of what might be causing that in a later chapter.

chance. As long as the birth rate and death rate stay the same, that growth will be exponential. Eventually exponential growth can lead to very large populations indeed. The residents of Medford witnessed that, literally. Yet, all biological populations have this power. Trouvelet's Gypsy Moths were only unusual in getting to express it so dramatically. The population of their London cousins, fortunately, has been more controlled in its growth.

The fact that we are not knee-deep in Gypsy Moths, or indeed in any species, shows us that growth rates do not stay the same forever. All populations eventually experience control. Either their birth rates fall, or their death rates rise, or both. When they reach equality, the population flattens out at a (more or less) stable size. Imperfect feedback from population size to growth rate can cause fluctuations as they overshoot and undershoot the level that the environment can support. Chaotic variation can result, even before the random environmental element is added. Stochasticity can easily take populations to extinction, as demonstrated by the Gypsy Moths of the English Fenlands. That tiny group of Gypsy Moths in Medford in 1868 or 1869 really got lucky. Léopold Trouvelet, and the people of Massachusetts, less so. As the moth population has grown and spread, it has defoliated millions of acres of New England forests, at huge financial cost. Almost $200 million was spent on monitoring and managing the "separate destroyer" between 1985 and 2004 alone.

There are huge elements of chance in the story of the Gypsy Moth in North America, but also much that follows basic ecological principles. The capacity for populations to grow exponentially underpins the field. A substantial portion of ecology involves us trying to understand the myriad ways in which they are prevented from doing so, and why population growth flattens off at higher abundances in some species than in others. In the broadest of terms, populations can be influenced from below by variation in the environment, but also by interactions with the species on which they depend for resources. Alternatively, they can be influenced from above, by interactions with the species for which they are themselves a resource.

It is to influences from below—to the importance of resources—that we turn in the next chapter.

Chapter 2

Footmen
The Consequences of Limited Resources

Hungry men may fight, but it will be for a bone. . .
—Frederick Upham Adams

Dingy Footman, Devon.

My first experiences of moth trapping in Scotland had quickly taught me that not all moths attracted to the light made it into the box. Indeed, the most interesting individuals were often to be found outside the trap. My morning routine after running the light has always involved a slow and cautious approach, with a thorough scan of the surrounding area. I dread stepping on one of the creatures I've lured in. This check is not an extended process on a London roof terrace measuring just a few square meters, but it's a rare day that I don't find a moth or two settled on the door frame or the wall. The latter in particular needs close inspection, as the browns and grays of several species give them near-perfect camouflage against the Victorian brickwork. It was at the base of that wall, on my first morning of moth trapping in London, that I found my first Footman.

A not insignificant part of the joy of catching and putting names to moths comes from the names themselves. They have been described as tiny poems, and many have a distinctly poetic quality. My favorites include Maiden's Blush, Flounced Rustic, True Lover's Knot, and the aptly named Merveille du Jour. The history and etymology of the names are fascinating topics in their own right.[i]

The origin of Footman denoting a group of species in the family Erebidae dates back at least to the eighteenth century, where it was first documented in *The Aurelian*, Moses Harris's classic book on Lepidoptera. The name was evidently inspired by the typical resting pose of these moths. Most Footmen sit with their grayish or yellowy wings tight to their bodies, looking like tiny stiff figures in formal tailcoat livery. Whoever coined their name had an artist's eye, and a sense of humor.

My Footman did not look like an employee of the highest society. His tails appeared unbuttoned and sloppily spread.[ii] They lacked the gloss and color of a well-cared-for suit. Reference to the books revealed that he was, appropriately, a Dingy Footman. Really not the best-dressed member of his family (although some individuals wear a more dapper

---

i. Peter Marren's *Emperors, Admirals and Chimney-Sweepers: The weird and wonderful names of butterflies and moths* (Beaminster, UK: Little Toller, 2019) is highly recommended.

ii. Or her tails.

yellow coat). Chris Manley's identification guide describes him as re-sembling a melon pip, and this is spot-on.[1] Still, he was exciting, being a new species to me. And as we know, names matter in science.

Three centuries on from *The Aurelian*, and liveried footmen are a declining breed in the UK. Happily, the reverse is true of their moth namesakes.

The Dingy Footman has one of the fastest-growing populations of any British species. In the years 1970 to 2016, catches from the national network of light traps run by the Rothamsted Insect Survey revealed an estimated increase of more than 5,500 percent in the British Dingy Footman population. Over the same period, data from the National Moth Recording Scheme showed that the extent of this species' distribution in Britain increased almost fourfold.[iii] While not a patch on the Gypsy Moth (with which it shares a family) in its early years in North America, this is still a spectacular increase. Dingy Footmen have spread rapidly across southern England and Wales since 1970, and are now in the process of colonizing Scotland and Ireland.

As it happens, the success of the Dingy is not exceptional among Footmen. British populations of most other species are also booming. The Common Footman was much more common in 2016 than in 1970—49 percent more, in fact. The Scarce Footman was much less scarce—its population increased 629 percent in the same period. Numbers of the Orange Footman have surpassed the Dingy, with growth topping 10,000 percent. Yet the Buff Footman puts them all in the shade. A 524 percent increase in the size of its British range between 1970 and 2016 has been accompanied by an estimated 84,589 percent increase in numbers. British Footmen are very much on the march.

Coming as they do at a time when (as I will explore in a later chapter) the general trend has been for British species to be in decline, these population increases beg an obvious question: Why have conditions

---

iii. The National Moth Recording Scheme is run by the charity Butterfly Conservation, in collaboration with MothsIreland.

apparently become so much more favorable for this particular group of moths? The answer can be found in their diet. Footman moths have caterpillars that feed primarily on lichens. The improvement in their fortunes is because of an improvement in the fortunes of their food.

I suspect that most of us think of lichens, when we think of them at all, as those little patches of gray-green or yellow that encrust gravestones or churchyard walls. This is to do them a great disservice. Lichens exhibit a wide diversity of forms—tufts of wild green hair, rows of reaching fingers, miniature red-flowered herbs, tiny ceremonial trumpets—and under a hand lens are as exquisite as any plant.

Lichens are not themselves plants, of course. Indeed, they are not even organisms in the traditional sense. Rather, they are an association between organisms—a symbiotic relationship between a fungus and a "photobiont." This latter partner is either a green alga (a type of plant) or, less commonly, a cyanobacterium, a type of bacteria that can photosynthesize (as do plants, albeit using different chemical processes).[iv] Plants are now thought to have started out the same way, with their photobiont becoming integrated into the host's cells as the organelle that we call the chloroplast.

The lichen's association is generally considered to be mutualistic—to the benefit of both partners. The photobiont brings the food to the table. It harvests sunlight, and uses this solar energy to process carbon dioxide and water into simple sugar molecules. Some of this sugar is then taken as food by the fungus, which needs sugars to grow and reproduce but cannot produce them itself. Fungi housing cyanobacteria may gain other nutrients from the ability of some of these photobionts also to fix nitrogen (capturing the gas from the atmosphere and transforming it into compounds that organisms can metabolize).

In return, the fungus provides protection, both physical and chemical. The body of the lichen is made up of fungal tissue, which offers a home for the photobiont, and shields it from desiccation and UV radiation, among other things. There is some debate over how much the photobiont really benefits from the fungus, or whether it is in fact just being farmed. (The same arguments can be made about the plants

---

iv. Some lichens have both algae and cyanobacteria.

in human agriculture.) Either way, the association is a widespread and successful one, with more than 25,000 lichen species known worldwide—though, as with moths, this number is likely to be a large underestimate of their true diversity.

Lichens have been a successful partnership over evolutionary history, but they have had a hard time in recent centuries, at least in industrialized regions. Lichens have no roots, and depend on "fallout" from the atmosphere to supply their water and those nutrients that the photobiont cannot provide. For this reason, they like it wet. While I mainly trap moths in London, I'm lucky also to be able to trap regularly in rural west Devon, at an old converted barn owned by my in-laws. The higher rainfall there often puts a damper on our family visits, but it is certainly to the liking of lichens, which adorn the trees with verdant tresses. The downside of this dependence on fallout is that lichens can be very sensitive to atmospheric impurities added by human enterprise. Air pollution is a serious problem for them. We have known about this for more than 200 years now, at least since observations made by the botanist William Borrer, in the early nineteenth century, that many lichen species could not persist in polluted urban areas.

Many chemicals are produced and released into the atmosphere as a side effect of industrial processes. Carbon dioxide is the one currently hogging the headlines (or it should be, at any rate). When I was young, it was sulfur dioxide. This gas is a by-product, among other things, of burning coal to generate electricity. It can dissolve into the droplets that make up clouds, reacting with the water to produce a dilute solution of that famously corrosive compound, sulfuric acid. These droplets can then coalesce and fall as "acid rain"—a familiar term to anyone who, like me, grew up in England in the 1970s and 80s. I remember television news back then running stories describing how British pollution was blown across the North Sea, to drop with fatal effect on Scandinavian trees.

Acid rain kills trees by dissolving key minerals out of their tissues, which then get washed away. The trees essentially starve. A similar process does for the lichen's photobionts, which cannot photosynthesize properly in acidic conditions. In fact, it's worse for the lichens. They absorb water directly, rather than via soil, and so the toxic pollutants accumulate more rapidly in their tissues. Lichens die and disappear while the trees they grew on still survive. They are therefore sensitive indicators of atmospheric pollution, and scientists have used them as such. The Industrial Revolution blackened buildings and trees across Britain's cities, driving the widespread disappearance of many lichen species from much of the country. Tall chimneys added to power stations to help vent pollution away from their neighborhood simply exported the problem. The natural world is one of myriad connections.

Yet the corner seems to have been turned for British lichens. Acid rain featured in the news reports of my childhood because the negative impacts of this form of pollution were well appreciated. Emissions of sulfur dioxide and other toxic gases from British industry were already in decline at that point. The UK was a party to the 1979 Convention on Long-Range Transboundary Air Pollution and subsequently signed up to a series of other international agreements aimed at cleaning up the atmosphere. The technology to remove gases from emissions produced by power stations and other sources had been available for decades by then. The shift from "dirty" coal to cleaner fuels for electricity generation also helped. As these elements of air pollution have dropped, the lichens have staged a revival. They do not swathe the trees in London as they do in Devon, but they are on the up.

And what's good for lichens is good for the caterpillars that consume them.

No species is an island, entire of itself. This was the elephant in the room in the previous chapter. I talked about constraints on the growth of a biological population without coming to grips with the primary necessity for that growth—the availability of resources. For animals, that means food.

All animals need something to eat at some stage in their life. (Plants need food, too, but these remarkable organisms make much of what they need themselves, and this book is more about the ecology of animals.) The presence of the right food is not in itself a guarantee that an animal that eats it will be present, but the reverse is more likely to be true. Nothing to consume means no consumers. Species can pitch up in areas without their food, but they will not stay for long. An army marches on its stomach.

When the size of an animal population depends on the availability of resources, it is said to be influenced from the *bottom up*. In the case of the Footmen, the lack of lichen seems to have been a key constraint on their populations. Loosening the constraint—in essence, raising the carrying capacity of the environment—can allow consumer populations the freedom to grow.

The spectacular growth and spread of the populations of British Footmen over recent decades is a testament to how the availability of food underpins the success of a consumer. They have flourished as cleaner air has allowed lichen populations to recover across much of industrial Britain. Not just Footmen, either.

As well as sixteen species of Footmen, another six species of macromoth resident in the UK have lichenivorous caterpillars.[v] We have enough information for twenty-one of these species to calculate changes in their distribution between the 1960s and 2010s. Twenty of the twenty-one species increased their distributions over that period. These include the stunning and well-named Marbled Beauty and the Tree-lichen Beauty, both of which are regulars in my moth trap through the height of the London summer. The latter is one of the more abundant species on the roof terrace. Yet it was only recorded in Britain for the fourth time in 1991, after three records in the nineteenth century. Its spread through southeast England has been as dramatic as that of the Footmen. That may not be down to the resurgence of lichen alone, but it has probably helped.

While the adults of most moth species do feed, primarily on nectar, it's the caterpillars that are the big eaters. Indeed, eating is pretty much the whole point of their existence, and they can put on substantial bulk.

---

v. We informally divide moth families into "macros" and "micros" —large and small. Macros are the species that have been the traditional focus of moth-ers, being generally easier to see and to identify than micros. This division has no scientific or biological basis, though, and indeed some micromoth species can be large and some macromoth species small. This can easily fool beginners in moth identification, as I'll demonstrate later.

Silkworms (caterpillars of the Silk Moth) may increase in mass 7,000 to 10,000 times between hatching and pupation—the equivalent of my daughter growing to the size of a Humpback Whale. A caterpillar's cuticle cannot expand to this degree, which is why they go through several larval instars, molting their old skin for a new one each time. The material they gain through all this consumption is reallocated in the pupa, where the caterpillar undergoes one of nature's most miraculous transformations to emerge days to months later as an adult moth.

The twenty species I find most often on the roof terrace account for almost three quarters of the moths that come to the London trap. By and large, the presence of an adult in the trap is a good indicator that its caterpillar has food in the neighborhood. The diets of their caterpillars paint a picture of the green space in the vicinity of a roof terrace in the London Borough of Camden.

The commonest species I catch is the Large Yellow Underwing, a moth named with functional directness. According to the field guide, its larval diet includes a wide range of herbaceous plants and grasses, including Docks, Marigolds, Foxgloves, and Brassicas (cabbages and their relatives—native species in the UK include many common, small, and disregarded weeds, such as Shepherd's Purse and Bitter-cress). The field guide authors must have created a keyboard shortcut for "a wide range of herbaceous plants," as it features for six of my top ten species; for a bit of variety, two more are described as feeding on "many" herbaceous plant species. Docks feature for several species, as do Foxglove, Dandelion, Common Nettle, and species of Plantain (the small native plants in the genus *Plantago*, not the large, starchy cooking bananas). In spring, caterpillars of several species climb up to feed on the leaves of larger and woodier species, such as bramble, Blackthorn (source of sloe berries), Hawthorn, and Sallow. The Lesser, Broad-bordered, and Lesser Broad-bordered Yellow Underwings all have this habit. The second commonest species on the roof terrace is the Pale Mottled Willow, whose caterpillars feed on grass seeds.

Moving down the list of common London moths, and a bit of variety is introduced. The tastes of the Box-tree Moth and Horse Chestnut Leaf-miner are largely self-explanatory. Caterpillars of the latter tunnel between the surfaces of the leaves, and the foliage of our local

Horse Chestnut is riddled with these mines. We have already met the Tree-lichen Beauty. I catch a lot of Grays—micromoths in the genus *Eudonia*, difficult for me to identify to species with certainty—which feed on moss. Caterpillars of the Ruddy Streak eat withered leaves and leaf litter. Those of the Double-striped Pug are catholic consumers of flowers, including Holly, Ivy, Gorse, Buddleia, and my neighbor's prized Roses. The Light Brown Apple Moth actually eats the foliage of many trees, shrubs, and herbaceous plants, although I'm sure that the (other) neighbor's apples must be the origin of some of those I catch.

Like the lichenivores, not all moth caterpillars eat plants. Some caterpillars can even be carnivores, consuming other caterpillars. Nevertheless, what the diets of the common moths of the Borough of Camden give us is primarily a list of plants, and plants common in rough ground, hedgerows, and parklands. In fact, to a significant degree the list describes the flora of nearby Hampstead Heath—an area that is a mosaic of sports fields and meadows, intersected with hedges, bramble patches, and clumps of more or less mature trees. Hawthorn and Blackthorn are common in the hedges. The herbaceous species are typical of disturbed areas—they live fast, and can quickly sprout, flower, set seed, and move on before shrubs and trees take over. On the Heath, open spaces managed for footballers, picnickers, and dog walkers provide large patches where the shrubs and trees never get a chance to take hold. Mowers now do the job, but not so long ago a flock of sheep grazed these slopes, with views over to Highgate, Westminster, and the City of London.

The key point is that the contents of the moth trap largely reflect the contents of the local parks and gardens. A similar story emerges from west Devon. I don't trap there as often as in London, but I do spend some time there every year. Once again, many of the common moths eat "a wide range of herbaceous plants," albeit that there is a bit more diversity in the species—knapweeds for the Treble Lines, Groundsel and bedstraws for the Flame Shoulder, and Wild Marjoram, Wood Sage, and Horseshoe Vetch for the Mullein Wave. The garden backs onto a field grazed more or less closely by bullocks, and moths with grass-feeding caterpillars make hay here—the Garden Grass Veneer is the commonest species in the trap, while the Dark Arches ranks fourth. The pastures are interwoven with tall, thick hedgerows and copses of mature deciduous trees. Tree-feeding species are much more common here than in London. The wild plum trees that supply me with fruit for making chutney also give me Brimstone Moths for the trap. There is a lot of oak and hazel in the mosaic, supplying Buff-tips, whose caterpillars feed on their leaves, and whose adults are masters of disguise. Someone nearby has a privet hedge, judging by how many Privet Hawk-moths I catch there in summer.

And finally, the abundant and luxuriant lichens of wet west Devon bring their reward in Footmen. Common is common there indeed, while Dingy and Rosy are regulars in the trap. Four-spotted, Orange, and Buff appear from time to time. So too do other species with lichenivorous caterpillars: Beautiful Hook-tip, Brussels Lace, and Marbled Green. The consumers depend on what there is to be consumed.

Recall that populations grow when births outnumber deaths, however small the advantage to births. We can say with confidence that this is true for those British Footmen whose populations are on a steep upward trajectory, like the Dingy and the Buff. But for no population do births maintain this advantage forever. Eventually, the birth rate will fall, or the death rate will rise, or both. And when these rates reach equality, population growth will stop. For most species, this results in their population reaching a more or less stable size. We've already seen how the logistic model of population growth gives us insight into the workings of this process. Food is fundamental to this equation.

In general, when food is abundant, a population of consumers can grow rapidly—essentially exponentially. Animals can eat their fill, and have plenty of energy and resources to devote to reproduction. This changes as numbers increase. Food supplies are necessarily finite on a finite planet, and eventually there will be too many animals trying to exploit too few resources. If an animal cannot find enough food to build eggs, or to provision a fetus, then it will not be able to reproduce. The birth rate will inevitably fall. If it cannot find enough food to maintain its own tissues, then it will weaken, and eventually die. The death rate will inevitably rise. Just as a key necessity for the growth of a population is a supply of food,[vi] so a key check to that growth is likely to be an increasing scarcity of supplies. An army marches on its stomach. Without provisions, the march grinds to a halt.

When a population reaches the point at which there are too many individuals fighting over too few resources, competition is the inevitable result.

Ecologists define competition as a negative interaction between organisms that require the same limited resource. It is negative because the *fitness* of one organism is lowered by the presence of another. Fitness here has the evolutionary meaning (sometimes referred to as *Darwinian fitness*) describing reproductive success, or an individual's contribution to the gene pool of the next generation. In lay terms, then, competition describes the fight to pass on your genes. All animals (indeed, all organisms) alive today can trace an unbroken line of descent through reproduction to the first life forms on this planet. Fail to reproduce, and that line is broken forever.

The logistic model of population growth is in fact implicitly a model of competition. In this case, the competition is *intraspecific*—that is, between members of the same species. It is the intensification of that intraspecific competition that eventually stops population growth: each individual added to the population reduces the resources available for others.

However, most animals are not conspecifics, but other species (the same is true for plants), and they may also have a say. Competition between species is termed *interspecific*. The fitness of one organism is still lowered by the presence of another, but now the organisms are members of different species. If two species share the same limited resource, competition between them becomes a real possibility. Recall the diets of the moths that come to my trap in London. Species after species had the same foods listed in their diets. Dandelions, Docks, and Nettles. Blackthorn and bramble. Think of all the species whose caterpillars eat lichens—and how fast their populations are growing right now. Opportunities for interspecific competition are rife.

Competition between species means that each of the opponents will negatively affect the population growth rate of the other, and depress the

---

vi. Or the equivalent resources necessary for growth and reproduction, in the case of plants and other autotrophs.

size of the population the other species can reach, given the resources available. Losing in intraspecific competition imperils individuals and their family lines. Losing in interspecific competition can put a whole species on the path to oblivion.

Species can compete in different ways.

Perhaps the most intuitive and common is exploitation competition, when competition is directly for a resource that two species share, but which can get used up. A morsel of food is the most obvious example. If the nuts in your bird feeder get eaten by squirrels, they cannot also feed the birds. Likewise the sloes in my gin. The resource is gone forever, to the benefit of one species and the detriment of another. If a Common Footman caterpillar eats a lichen, that lichen is not available for a Dingy Footman caterpillar, should one crawl down the same twig. The same plant species feature in the diets of many of the commonest species in the moth trap. It's likely that at least some of their caterpillars will be competing to exploit those plants.

Species can also compete through interference. Here, one species doesn't directly deplete the resource, but instead prevents access by competitors. The classic example of interference competition is territoriality in birds—an individual, pair, or family will aggressively defend an area to monopolize the resources it encompasses. Territories can be guarded to keep out other species, as well as their own. Populations of Rice Moth and Almond Moth will persist perfectly well on cocoa beans in laboratory cultures if kept separately, but Almond Moths die out if the species are housed together. This is at least partly because the larger and more aggressive Rice Moth caterpillars monopolize pupation sites. A more extreme form of interference is to kill competitors. Leaf-mining moth caterpillars are known to do this when only one individual can develop on a leaf. This certainly interferes with the ability of a rival to feed, and without depleting the available resource.

Species can also compete through preemption. Now competition is for a limited resource that two species share, but which does not get depleted by the species that gets to use it—it could in theory become available for another species at a later date. Space is the classic example here. Nest holes are a limited resource for many bird species—which is why nest boxes are good additions to gardens and nature reserves—and holes can only be recycled once any incumbents have fledged and left. The trees turned yellow with Gypsy Moth eggs in Medford may well have been preempted as laying surfaces, to the detriment of other species.

However species compete, the result is that the population sizes of each are reduced. Resources that would have been available for a species given free rein are no longer available once a second species is using them. Species inevitably lose out in competition. But what does "losing" mean in reality? What happens to a population of one species when another comes on the scene—does it decrease, or does it die out altogether? Do all species lose to some degree, or can there be winners? The answers—as will frequently be the case through this book—are that "it depends."

To understand *why* it depends, we start with the model for logistic growth from the last chapter. Instead of thinking about how density dependence affects births and deaths in just one species, we now have to factor in the additional effect of a competitor. For simplicity I'll just consider two species, but in theory more species could be involved. Even with just two species, though, a variety of different outcomes for the populations of both are possible.

The classic model for interspecific competition in ecology was worked out more or less simultaneously by two scientists approaching the question from quite different perspectives—Alfred Lotka and Vito Volterra. Lotka was an American chemist and mathematician whose contributions to ecology came from his interest in applying equations from chemical systems to biological ones. Vito Volterra was an Italian mathematician and physicist. Volterra was inspired to apply mathematical ideas to biology through his relationship to a marine biologist, Umberto D'Ancona, who was studying fish catches in the Adriatic, but more importantly, was courting (and later married) Volterra's daughter. Career paths can take unexpected turns. Alfred and Vito's names are now indelibly linked. The Lotka-Volterra equations are the base on which ecological models of competitive interactions are built.[vii]

---

vii. And, as we'll see later, those between predator and prey.

The Lotka-Volterra model assumes that both populations would normally follow the classic *S*-shaped logistic growth pattern if each population were growing alone. Each individual added to it reduces the amount of food available, with knock-on incremental decreases in birth rate and increases in death rate.

The arrival of a competitor species simply introduces another drain on the food supply. Now, the amount of food available depends on the populations of both species. The population growth rate of each species decreases not only because it is adding individuals to its own population, but because its competitor is doing the same. The presence of a competitor lowers the carrying capacity of the environment for a species by lowering the amount of food available—birth and death rates equalize at a lower population size. It's obvious, really—a species is not going to be so common if another species is taking away some of the resources it relies upon.

The question then becomes: How large is the population of each competitor species going to be, once competition has run its course?

The answer depends on how good each species is in competition, both with the other, but also with itself. The Lotka-Volterra model has a *competition coefficient* to quantify this for each species. The competition coefficient tells you whether the population growth rate of a species is going to be reduced more if another individual of its own species is added, or if that extra individual is one of the rival species. If the competitor's coefficient is less than 1, then competition is fiercer within the species than from the rival: intraspecific competition is stronger than interspecific competition. If it's greater than 1, then interspecific competition is a stronger dampener than intraspecific: each individual of the rival species reduces available resources more than each extra kin.

The bottom line is this: who wins in competition, and indeed whether or not there *is* a winner, depends on whether intraspecific or interspecific competition matters more for each species. With two species, the Lotka-Volterra model has four possible outcomes. I'll illustrate this with the hypothetical case of two competing Footman species, the Buff and the Dingy.

If intraspecific competition has a stronger effect on the Buff Footman than does competition by the Dingy, then the Dingy is clearly a weak interspecific competitor. Each extra Buff has more of an impact on the Buff population than does each extra Dingy. If each extra Buff also has more of an impact on the Dingy population than does each extra Dingy, then the Buff is a strong interspecific competitor as well. Strong beats weak in interspecific competition, and the Dingy Footman population goes extinct.

The reverse may instead be true, with the Dingy Footman the strong competitor, and the Buff Footman the weak. Now, the Dingy wins, and the Buff population goes extinct.

However, there can be situations where both species are weak interspecific competitors. Each extra Buff Footman has a greater dampening effect on its own population growth rate than does each extra Dingy, and likewise for the effects of each extra Dingy and Buff Footman on the Dingy population. Intraspecific competition matters more than interspecific for *both* species. Now, both species can coexist. Moreover, the coexistence is *stable*—because both species are weak interspecific competitors, neither can push the other species to extinction. If the Dingy Footman population by chance declines a little, it can always bounce back, because it is a stronger competitor against itself than is the Buff (and vice versa).

The fourth outcome is where both species are strong interspecific competitors. The Dingy Footman is a better competitor against the Buff Footman than the Buff is against itself, but the same is true for the Buff against the Dingy. Interspecific competition is stronger than intraspecific competition for both species. In this situation, it is *possible* for the two species to coexist, but it would be a delicate balance, and coexistence is now unstable. If one species gets the upper hand in terms of numbers, it quickly pushes the other population down to extinction. Each extra Buff Footman would drive a further nail in the coffin of the Dingy, and vice versa, through the stronger effect of interspecific than intraspecific competition.

So, when two species square up against each other for a limited resource, the outcome can be win, lose, or draw. Which outcome we expect will depend, and Lotka and Volterra's model helps to clarify upon what.

To me, competition between species has always seemed rather a weak tool for controlling populations. It's true that the same food species appear repeatedly in the diets of the commoner moths I trap on my London roof terrace—those herbaceous plants in particular. The overlap in diets would undoubtedly be greater still were I to extend that list beyond the top twenty species I catch. Yet there is no shortage of those plants, even in this urban London borough, and don't forget the wide variety of them consumed by the caterpillars. Have you ever bent down to a Foxglove, Nettle, or Dandelion plant and found caterpillars of even one moth species, let alone two? I'll admit that you might, but if you have, I bet you had to search over quite a few leaves, if not plants. Does competition really matter that much in the real world?

Well, yes, it does. Even when the food being fought over is larger than a Dandelion. Imagine the impacts that the Gypsy Moths in Medford must have had. There can be no doubt that stripping whole groves of their leaves will have decreased the fitness of other tree-feeding moth species. This is how we define competition. We can see these effects in moths here in England, too.

Wytham Woods has for many years been home to field studies by ecologists and evolutionary biologists based at the nearby University of Oxford, which has owned the site since the 1940s. It's particularly famous for research on its bird populations, especially the Great and Blue Tits. To understand the birds, ecologists have had to study other species, too—the insects on which the birds depend for their food, and the plants on which those insects depend for theirs. Some of the most important studies concern a pair of moth species—the Winter Moth and the Green Tortrix.

In some years, caterpillars of these two species can attain densities high enough to completely defoliate individuals of the food plant they share. We're not talking about small herbs now, but that most iconic of British trees, the Pedunculate Oak. These are not alien species seizing a new opportunity unchecked, like the Gypsy Moth in America, but native species consuming in their normal environment. Even in average years, the Winter Moth and Tortrix together can remove 40 percent of the leaf area of trees in Wytham Woods. It's an impact that can be read in the very bodies of the trees: comparing levels of defoliation with the thickness of tree rings over the years suggests that oaks could accumulate 60 percent more wood in summer were it not for the caterpillars.

So here we have a pair of moth species feeding on a shared, finite, depletable food—oak leaves. Caterpillars of both species can be found on the same leaf buds and clusters. Consumption of the leaves is so high that they ought to be in direct competition. And indeed, the evidence is that they are.

Variation in the size of the Tortrix population in Wytham is most strongly determined by the availability of food, through intraspecific competition between their caterpillars for leaves. Tortrix caterpillars survive better when they are the only species present in clusters of oak leaves than if Winter Moth caterpillars are also present. On the other hand, Winter Moth caterpillars do better when mixed with Tortrix caterpillars than when the competitors are of their own species. This is all consistent with Winter Moth caterpillars being strong competitors, and Tortrix caterpillars being weak.

Some of the effect of the Winter Moth on the Green Tortrix seems to come through interference competition. The Tortrix is a "leaf roller." Its caterpillars roll or fold leaves around themselves, and feed and grow in a moist, protective, oaky tent. If a Winter Moth caterpillar nibbles holes in this wrap, though, it can ruin the Tortrix's homemade microclimate. The resident may dehydrate. Damaged leaves are also harder to roll. None of this is good for Tortrix caterpillars.

Whether through interference or exploitation, the Winter Moth seems to have the upper hand in this particular competition. But according to Lotka and Volterra, the strong competitor should oust the weak. Why then is the Green Tortrix not driven out of the woods?

The answer is probably that oak leaves are not a limited resource in Wytham Woods in most years, even given the high levels of defoliation by the moths. Simply, there's usually too much food for competition to drive out the Tortrix completely. Competition is only likely to kick in seriously on branches where caterpillar densities are high, or those occasional years when defoliation is unusually severe. On the whole, oak leaves don't often get rare enough for competition to imperil the Tortrix population, and even when they do, they don't stay rare for long.

Adding another layer of complexity, moths don't only compete with other moths.

In Britain, Ragwort is a familiar plant of waste ground and grazing pastures. It is often loathed by landowners, as it's toxic to cattle and horses. Ecologists love it, though, because of the diverse community of insects and other invertebrates it supports. Perhaps the best known of these is the Cinnabar Moth. Ragwort plants are frequently draped with its tiger-striped caterpillars. The adult moth is also a brightly colored beauty, with elegant gray-and-red upperwings and bright red underwings. It's one of the relatively few day-flying British moths,[viii] and a common sight around the streets of Camden. The relationship between moth and plant has been well studied. Ragwort is an alien weed in Australia and Canada, and the moth was a candidate species to release for biological control (the agricultural equivalent of swallowing a spider to catch a fly, albeit ideally from a more rigorous scientific footing). In such situations, it's important to know the relationship between the weed and its potential nemesis inside out.

Cinnabar Moth caterpillars can chew ragwort plants to their stems, but it turns out that the direction of control is not plant by caterpillar, but the reverse. Ragwort abundance depends on rainfall at the point when seedlings are establishing, with plant abundance one year predicting Cinnabar Moth abundance the next. The moths depend on the food plant, but are passengers on changes in Ragwort abundance, not their driver.[ix] This is a classic bottom-up effect—consumers depend on the food they consume. The Cinnabars can gorge themselves when it's good times for the plant, but famine can quickly follow feast.

Consumption by Cinnabar Moth caterpillars does not affect plant population growth, but can impact populations of other consumers. One such is the Ragwort Seed (or Seed-head) Fly. You can probably guess how this species makes a living, but for the record, female flies

---

viii. Although there are more species of day-flying moths than there are butterfly species in Britain.

ix. Information like this is *why* it's so important to know the relationship between the weed and its putative control agent inside out. The moth would probably have had no effect on the weed population in this case, wasting a lot of time, money, and effort—and with the potential for unintended consequences for other species to boot.

lay their eggs in Ragwort flower heads in late spring and early summer, and the larvae that hatch from them eat the seed heads and destroy the seeds. Unfortunately for the fly, the moth is also partial to Ragwort flower heads. If caterpillars are present on a plant, they can cause the deaths of all the fly larvae by classic exploitation competition. The fly has no comeback on the caterpillar population. The moth is the strong competitor, and the fly the weak. As with the Winter Moth and Green Tortrix, though, exactly how this contest plays out depends on how much food there is to go around. The flies are not impacted by the moths when there are flower heads in abundance. Recall the definition of competition—a negative interaction between organisms that require the same limited resource. When resources are less limited, competition can be less fierce. Spread the table and contention will cease.

So competition is a force that matters to plant-eating insects, such as the moths that come to my trap. When food is limited, populations of consumers are limited as well, bottom-up. And when different consumers depend on the same resource, as is the case for many common British moth species, the stage is set for interspecific competition. Losing can be the end not only of individuals, but potentially of whole populations. Weak competitors will lose out to strong. There is not even security in strength, because strong competitors can still lose out to other strong competitors, as the Lotka-Volterra model shows.

Yet, finding examples where competition alone causes the loss of one of the competitors is surprisingly difficult. Even classic examples, like the retreat of the native Red Squirrel from most of Britain in the face of the more competitive alien Gray Squirrel, usually turn out to be less straightforward. For the squirrels, a disease carried by the Grays and fatal to the Reds probably plays a—perhaps *the*—major role. Competition certainly *is* a force in nature, but perhaps it's not as detrimental as the models make out. What's going on?

The answer is that competition provides serious motivation for species to find ways to live together. Because competition can be fatal for a weak (and even a strong) competitor, species find ways to separate

64 The Jewel Box

out how they use the environment. This is known as *niche differentiation*.

The niche describes the set of conditions that a species requires to persist. It includes both the physical environment and biological necessities, such as food and other resources.[x] It is a maxim in ecology that if two species have identical (or very similar) niches, then whichever is better at using that niche will inevitably drive the other to extinction—the *competitive exclusion principle*. No two species can live in exactly the same way. If they do, strong beats weak. If true, then there must be a limit to how similar the niches of two species can be if both are to persist. Differentiating the niche reduces its similarity to a close competitor, and allows species to coexist without competing. In Lotka-Volterra terms, a species must shift to parts of the environment where their opponent is a weak competitor.

Necessity is the mother of invention, and species have been inventive at differentiating niches. One approach is to move apart in space. There are many different parts of a plant for a phytophage to eat, even on small plants like grasses. Garden Grass Veneer and Pale Mottled Willow are common moths in my trap in London—the Veneer's caterpillars feed at the bases of grass stems, those of the Willow up on the seeds. My neighbor's roses may have Double-striped Pugs on the flowers and Common Emerald on the leaves. Apple Leaf Miners tunnel between the surfaces of leaves, while Codling Moth caterpillars tunnel into the apples themselves—local gardens supply both species to my trap in London and Devon. Caterpillars of the several species of swift moths (in the family Hepialidae) live in the soil and feed on roots—of grasses, bracken, and herbaceous plants, depending on the species.

Competitors can also divide up space by focusing on different plant individuals. The Ragwort Seed-head Fly loses out to the Cinnabar Moth when larvae of both are feeding on the same plant, but the fly distances itself from competition in a literal sense, finding refuge in small, isolated clumps of Ragwort plants, too small to support a population of Cinnabar Moths, and too far away from other plants for caterpillars to reach by crawling, or adult females (which are quite sedentary) by flying.

---

x. We will hear more about this in a later chapter.

Competitors can also slice up their environment in time. Winter Moth and Green Tortrix caterpillars largely feed on oak in the spring, as do the caterpillars of most other leaf-chewing moths. Oak leaves are at their most tender and nutritious then, soft and proteinaceous. Chewing caterpillars try to make the most of this spring glut; older leaves are tougher and loaded with unpalatable tannins. In contrast, leaf-mining species on oaks tend to feed in summer, when competition with chewers is less. Damage from chewing stimulates oak leaves to produce more protective chemicals, increasing the death rate of leaf-tunneling caterpillars, which, unlike chewers, can't escape to another leaf. Miners avoid these impacts by waiting until the chewers have eaten their fill. Of course, this can be a dangerous strategy if the chewers' fill includes every leaf, or enough that pickings are slim from what's left. It's usually best to get to the buffet early, if possible.

It's an irony of competition that it is a force we often see in its absence. When we find two species that we think ought to be competing—eating the same food plant, for instance—making slightly different livings, it's easy to leap to the conclusion that competition is in play. When species are apparently avoiding each other, we can invoke the "ghost of competition past" to explain it: species are not competing now but they used to, before they differentiated their niches. Lots of patterns we see in nature are consistent with competition, past or present, but do they really result from it? It's easy to tell Just So stories. The reverse might also be true—we overlook competition when it's there, because it forces species apart to the point that we cannot see its effects.

Observing nature is important, but this shows why it is only the start of the process of understanding. We observe, and then invent explanations—hypotheses—to explain our observations. Our hypotheses may be verbal models, or they may be mathematical models of the sort that Alfred Lotka and Vito Volterra proposed. We then have to test whether these hypotheses are correct. Usually, we manipulate the system we are interested in, in ways that we predict will reveal its workings: we experiment. It is only through these experiments that we can truly reveal the workings of competition.

Flies and moths live on different Ragwort plants, but is this an effect of competition? What happens if we remove the caterpillars? Doing this experiment shows that the flies actually do better on the plants with the caterpillars removed. The experiment tells us what simple observation cannot—that competition was keeping the two species apart in space. Forcing leaf-mining moths to lay their eggs in spring shows that they grow and survive better on spring leaves than on the summer leaves they normally mine—unless those leaves are damaged by chewing caterpillars. Experiments show that competition drives miners and chewers to separate their activities in time.

It is the fact that science tests its ideas, and rejects those ideas that fail the tests, that sets it apart from other approaches to seeking truth. Even when our models are wrong, they can still be useful by helping us to understand *why* they fail. Other models can then take their place. Ideas win and lose in competition, too. Either way, our understanding grows.

In most cases, though, we still don't know if or how competition is happening. The Footman caterpillars all feed on lichens. It's difficult from knowledge of their natural history to see how they might differentiate their use of this resource. There are thousands of moth species within the limited borders of the UK alone, and any one of them could be competing with many other species, not just other moths. Thousands of species, millions of possible interactions. Experiments require *a lot* of painstaking work to do properly. Most of our ideas about how these species compete, or might have competed in the past, are still just ideas. At least we know that competition *does* matter for a handful of our moths. The concept has been proved. For now, that will have to do.

Resources matter, and for an animal, the most basic of these are food and water. The latter is not a problem for British moths, on the whole, and so food is the primary determinant of what I find in my trap. If caterpillars can't eat enough to make the transformation into adults, then there will be no adults. Walking along my local streets and up on to Hampstead Heath, I can see in broad terms what grasses and herbaceous plants are common, and what shrubs and trees overshadow them. From that, I can start to imagine what I might expect to find in the trap, and what I probably won't. From day one I had gazed in hope at the tall Limes at the bottom of the gardens overlooked by our terrace. It was a glorious May morning in 2019 when I finally opened the trap to find a Lime Hawk-moth.

Populations will grow as long as their basic needs are met, and in times of plenty, they can grow very quickly indeed. The spectacular upward trends in abundance and distribution shown by several British Footman species at the moment suggests that the lichen harvest is currently plentiful. In a period overshadowed by the threat of climate change, it is a tangible vindication of international treaties to clean up our environment—they can work, if we have the will to make them work.

However, one does not have to be Nostradamus to know that the Footman populations will not climb upwards forever. Good times always come to an end, and at that point the knives come out. Every society is only three meals away from chaos. We are the only animal with conscious awareness of the fact that we live on a finite planet, not that our behavior generally evidences it.

When rations are short, life quickly becomes a struggle for survival. This makes things difficult for individuals in a population, but as soon as populations of multiple species are involved, whole species can be at risk. Weak competitors can lose out to strong ones. So too can strong competitors—there is no safety in strength alone. Species have to find ways to live where their opposition is weak, so that the greater struggle is with their own kind—that is, where intraspecific competition matters more to them than does interspecific competition. By moving their niches apart, in space or time, competing species can create enough distance between them that both can persist.

Resources and how species use them are important for the moth trap, but understanding that is just the start. Resources determine the sorts of species we catch, but what about how many species? If we plant greater diversity into our gardens, will we get more moth species, or will there not be enough food of each type to support the populations we already have? If competition determines how similar species can be and persist, does that also affect how many different species we can end up with? And given that moths have to fly to get to the trap, how far are they coming?

These are questions that will concern us in later chapters, but there is a more manifest one to consider. Most caterpillars eat plants, or other organisms, like lichen, that depend on photosynthesis. This makes eminent good sense. To terrestrial creatures like ourselves, the natural world is a sea of green—a sea of caterpillar food. Yet finding caterpillars is generally hard work. You'd have thought, given all this food, that there would be a lot more consumers. Why aren't there?

One answer to this is that all species try to find ways to defend themselves against being consumed. Plants and other types of caterpillar food are no different. However, another answer stems from the fact that the consumers can themselves be the consumed. It is to this sort of interaction that we turn our attention in the next chapter.

Chapter 3

The Oak Eggar
When Consumers Become the Consumed

We behold the face of nature bright with gladness, we often see superabundance of food; we do not see, or we forget, that the birds which are idly singing round us mostly live on insects or seeds, and are thus constantly destroying life. . . .
— Charles Darwin

Oak Eggar, Devon.

S uccessful science is very much about prediction. We want to reach such a level of understanding about our study system that we can accurately forecast how it will behave in the future. How it will respond to disruption. How a similar system will act.

When we get this right, the effects can be spectacular—indistinguishable from magic, to paraphrase Arthur C. Clarke.[i] Physicists are masters of the art. The theory of quantum electrodynamics predicts properties of the electron with an accuracy, according to Richard Feynman, equivalent to estimating the distance between New York and Los Angeles to within the width of a human hair. I write this just days after NASA has delivered a one-ton vehicle to a predetermined location on Mars, after a journey of nearly 300 million miles It is nothing short of miraculous.

Against that, ecology sometimes gets a bad rap. Attempting to predict the behavior of living systems with such precision just makes ecologists look bad. Scientific successes like the discovery of the Higgs Boson simply serve to remind us that we still don't know perhaps 80 percent of the species with which we share our planet. Or perhaps 98 percent—we don't really even know how much we don't know. We share the planet with trillions of other organisms. Trying to work out the fundamental interactions between them, and how those interactions generate structure and pattern in diversity across the globe, is hard enough anyway. It's harder still when we don't even have names for most of them. It gives ecologists physics-envy.

It was partly for this reason that I was pleased when, one overcast August morning in Devon, I extracted two Oak Eggars from the moth trap. I had predicted their appearance more than a month earlier (not to the precise day, I'll admit) and so could not help feeling a little smug to have that prediction fulfilled. In any case, on top of my self-satisfaction, Oak Eggars are glorious moths to catch. They are among the larger British species, as large as Hawk-moths, with broad wings and bodies luxuriously coated in fur. The two I caught were both females, the larger sex, pale buffy-brown with a dark-rimmed white spot and broad yellow

---

i. One of Clarke's "Three Laws": "Any sufficiently advanced technology is indistinguishable from magic."

band on the forewings. The overall look is not unlike a bemused Honey Monster.[ii]

A female moth can typically lay hundreds of eggs over her lifetime, with females of some species laying on the order of 20,000 or more. The population growth that can be engendered even by species at the lower end of this range of fecundities was well illustrated by the Gypsy Moths in Medford. Populations will grow as long as just over two new individuals, on average, are added to a population for every female, at least for animals that reproduce sexually. Only two offspring, from the hundreds that a moth can produce, need to survive for a population to persist at a steady level. As discussed in the previous chapters, the fact that we are not knee-deep in moths shows that most populations experience control—numbers are replaced from generation to generation, but in general do not grow substantially. It follows that the embryos in most of the eggs laid by a female moth do not end up producing eggs of their own—which is, with some rare exceptions, their only reason for existing. Most moth lives end unfulfilled.

That adds up to a lot of death. We have met some of its causes already.

Environmental stochasticity can erase whole populations when times are especially harsh, especially if those populations are small to start with. A competitor can deprive another animal of essential resources, especially when those resources are limited. It can be difficult to eat at a table alongside 200 hungry mouths, let alone 20,000. Some caterpillars deal with sibling rivalry in brutal fashion, chowing down on unhatched eggs containing their potential siblings. They boost their own chances of survival and eliminate the competition in one easy meal.[iii] That is competition within a species. Competition *between* species can be serious enough that they each have to find a time or a place to live where (or when) they are better than the rest, or the whole species can

---

ii. That is, the fluffy, bearlike breakfast cereal mascot.

iii. In fact, sometimes some eggs are laid just to *be* eaten.

be starved out. Many deaths may happen this way, especially when caterpillar food is not as abundant for a species as it might seem from the sea of green with which we are surrounded.

Yet many deaths also happen because animals not only consume other species, but are also consumed by others. Even in the artificial confines of a moth trap, it's impossible to ignore the truth that nature is red in tooth and claw. The trap not only reveals patterns in diversity, but also some of the controls upon it. I had predicted that I would trap an Oak Eggar because it is not only moths that are drawn to the light. Indeed, the morning after a trapping night can be quite stressful. The trap not only attracts moths, but their consumers, too.

In Devon, the moth trap is a big draw for the local Robins. In 2020, a pair bred in a hole in the yard wall, and the strain of raising a brood during a national lockdown was all too evident on the female. She would sit in ragged plumage, watching me as I worked the trap, swooping on any moth that managed to elude my grasp. As her brood grew, her feathers became increasingly tatty, and her sallies bolder. Sometimes she would hop to within touching distance to grab moths off egg cartons I'd removed from the trap.[iv] Torn between twin loves of birds and moths, I didn't feel too guilty on those occasions when I failed to repel her advances. It's not like she'd have been a vegetarian otherwise. The resident Pied Wagtails also learned that early-morning visits to the garden could be a useful source of insects, and the three young they fledged from under our eaves that year were partly reared on escapees from the trap. Great Tits were the boldest of all, perching on the trap if I strayed too far. Sometimes their search for a meal even took them inside it.

Tits are prodigious consumers of moths. As we saw in the introduction, moths form the bulk of the more than two billion caterpillars estimated to be fed to nestlings of just the Great Tit and Blue Tit in the UK

---

iv. Empty egg cartons are put into the trap to add internal structure and provide resting places for the moths.

every day in the breeding season. Broods of both species take around three weeks to mature, and parents can raise more than one brood in a summer. The adults themselves need to feed too, of course—that could easily sum to more than fifty billion moths consumed each year, just by these two species alone, and just in the UK. There are four other tit species resident on these islands, and while none of them are as abundant as the Great or Blue, all feed nestlings a diet in which Lepidopteran caterpillars figure highly. Tits are a family in the order Passeriformes, and most of these birds include invertebrates in their diet in the breeding season—even species like larks and finches that are typically viewed as seed-eaters. That's more than 124 million parent passerine birds in the UK alone, feeding themselves and their broods at least in part on moths. More than 80 percent of bird species worldwide eat invertebrates, and the various life stages of moths will figure in the diets of many of these. I try not to get too worked up about the odd moth that gets pinched from the trap.

Birds have good eyesight, and they use this to good effect in searching out moths at all stages in their life cycle. The death they bring has forced moths to respond. One response has been to try to become invisible. Many species have cryptic coloration as adults or immatures. Browns and greens predominate in moth field guides, designed for camouflage against wood or leaf. The particular color reflects the preferences of the species for resting sites—the rich greens of the Emerald moths help them dissolve into foliage but are less helpful when the trap attracts them to the Victorian brickwork of my roof terrace. Even strikingly patterned species can disappear against the right background. The stunning Black Arches—large, largely white, but with black lines (like the feathered icing on a Bakewell tart) that give it its English name—can all but vanish when at rest on a lichen-covered rock or branch. The adult Buff-tip is a ringer for a piece of broken birch twig, and a favorite of many moth-ers. Many caterpillars are astounding stem mimics, but others take camouflage to the next level by covering themselves with small pieces of twig or bark. Bagworm moths live in silken pouches so adorned, sometimes even as adults.

Becoming invisible is a good strategy against a visual predator, but birds can still smell the pheromones released by caterpillars, and find them that way. Plants also release chemicals to attract birds to chewed leaves in order to speed caterpillar removal.

There's more than one way to skin a cat, though. Moths can take the opposite route and become obvious. Some species eschew camouflage for warning: being brightly colored usually signals that a species is unpalatable. Many feed as caterpillars on plants with toxic defensive chemicals, and have evolved to store these toxins in their tissues. The tiger-striped caterpillars of the Cinnabar Moth make no attempt to hide from visual predators, but instead advertise that they have laced their bodies with toxic alkaloids co-opted from their Ragwort food plant. They should not be eaten. The black-and-red adults send the same message. Some moths make themselves obvious *as* camouflage. Several species disguise themselves as bird droppings produced by large (the Scorched Carpet) or small (Chinese Character) bird species. On occasion I have almost overlooked Chinese Characters sitting on the moth trap, assuming them to be presents left by visiting Great Tits. Still other moths use bright colors and crypsis in combination. The apogee of this in Britain is the stunning Eyed Hawk-moth. Its camouflaged upperwings mask brightly colored orange underwings, in which are set the image of a pair of bright blue eyes. The moth flashes these eyes to startle predators.

Most birds are diurnal. Once the sun has set, their place as winged vertebrate predators of moths is largely taken by bats. There is a bat loft in the converted barn in Devon, and dusk in summer often signals the appearance of Pipistrelles fluttering around the eaves and hedgerows. The moths lured to my trap do not have much to fear from the tiny mammal, which mainly feeds on less substantial insects such as swarming midges and other Diptera (flies). However, Britain is home to eighteen bat species, and many of these eat adult moths. Worldwide, bats are second only to rodents among mammals in terms of species numbers, and most of the 1,400 or so known species are insectivores. The nocturnal habits of most bats and adult moths inevitably pit one against the other.

The importance of bats as predators is illustrated by the remarkable adaptations that help moths to evade them. Bats locate moths and other flying insect prey using echolocation—natural sonar. They produce streams of high-frequency ultrasound that reflect off objects such as vegetation or prey and bounce back to the bat's sensitive ears. Patterns in the reflections give information on the size and location of objects, allowing bats to navigate through their environment, catch prey on the wing, and even pick perched insects off foliage. In response, moths have developed sensitive ears (located on all parts of their bodies), so that they can hear bats coming and instigate evasive maneuvers. A wide range of species can produce their own ultrasound, which serves to jam echolocation, startle the predator, or even communicate that the moth may be unpalatable. Some moths seem to have modified their wing scales in defense, evolving structures that absorb bats' ultrasound. The scales act as acoustic "cloaking devices" to mask the insect from their predator's senses. The fur on their bodies does the same job. These adaptations would not have appeared without the pressure of natural selection, which gives advantages in survival and reproduction to moths that can better evade hunting bats.

Birds are not the only predators drawn to the moth trap. Late summer and autumn mornings often dawn with wasps sitting torpid on the egg cartons. Sometimes I find their larger cousins, hornets. These huge yellow-and-rufous insects have made me nervous ever since an unfortunate walk in the nearby woods, where my young daughter was stung after somehow ending up with two in her hair. I always need a few deep breaths before I ease them into a collecting tube for relocation.

The moths have more cause to be nervous than me. Wasps are important predators on other insects in general, and once they have roused themselves will frequently attempt to grab a moth. On one particularly memorable morning in London I caught some movement out of the corner of my eye, and looked up to see a wasp barreling up to the trap. It dropped straight onto a Box-tree Moth resting on one of the egg cartons, kicking off a frantic wrestling match between the two. The moth's attempts to shake off the wasp flipped them both around the table, then off the edge and down to the terrace floor. The jolt of landing gave the moth a moment of respite, but the wasp was quickly back on the attack. A jab of its sting and the moth's resistance began to ebb away. The fight quickly became one-sided as the toxin took effect, giving the wasp leisure to bite off the moth's wings and legs, leaving just the fat and protein-packed body to carry back to its nest. The wasp glittered with silvery moth scales that had come loose in the struggle as it flew away with its streamlined parcel clasped firmly in its mandibles. I was left slack-jawed, standing on a terrace littered with discarded moth appendages.

I've assumed that talk of wasps in the previous paragraphs has had you imagining the well-known yellowjackets, and if so, you would have been right. In that context, at least. Yellowjackets and hornets are Vespids, members of a family of around 5,000 species worldwide in the Hymenoptera, the insect order that also includes ants and bees. Like honeybees, yellowjackets are eusocial: they live in colonies founded by a single reproductive queen, served by workers who are also her daughters. One of the key worker tasks is to bring food back to the nest for their larval siblings (sisters mainly), and this was what brought a wasp down on the Box-tree Moth at my trap. Yet, in all of these characteristics, yellowjackets are relatively unusual as wasps go. Most wasps are not Vespids, not eusocial, and not nest builders. Most are not predators, in the colloquial use of this term. Most wasps are parasitoids.

Parasitoids are the stuff of insect nightmares. Halfway between predators and parasites, they lay their eggs near, on, or even in other insects, and the larvae that hatch out then live by slowly eating alive an unfortunate host. Think Ridley Scott's *Alien* and you will get the general idea.

Not all parasitoids are Hymenoptera (some are Diptera, flies mainly in the family Tachinidae), but the great majority are—either Ichneumonoidea (the superfamily containing most of the larger species) or Chalcidoidea (most of the smaller species). Parasitoid Hymenoptera attack the full range of insect life stages, from egg to adult (and other arthropods too, notably spiders), but most feed in or on larvae or pupae. None depends on adult moths, though, for reasons that are not entirely clear.

Some parasitoids, termed *idiobionts*, permanently paralyze their hosts. This is a common strategy when the wasp larvae develop outside their hosts, as paralyzed hosts cannot then dislodge them. Those that develop inside host bodies—*koinobionts*—usually let their hosts continue to develop and grow. They benefit in terms of food, because the host continues to eat, and in terms of protection, because the host can still evade predators. The end point is always the death of the host, though, as the wasp larva—or larvae, as hundreds of some species can develop inside a single host—completes its growth and consumption. The parasitoid larva may then pupate within the empty skin of the host, or eat its way out, like the Alien from John Hurt, to pupate elsewhere.

Parasitoid wasps include the smallest insects known to science. These are the so-called fairyflies that develop in the eggs of other insects.[v] Fully-grown adults may be less than a hundredth of an inch from head to tail. Other parasitoids can reach impressive sizes, though. It was the appearance of one of the largest British species in the moth trap one June morning that had led me to my prediction of Oak Eggars.

Enicospilus inflexus is an ichneumonid wasp, one of nine species in its genus found in the UK. All are large, orange insects, with long, slender abdomens extending back from the narrow waist that is one of the defining characteristics of the Hymenoptera. *E. inflexus* is large relative even to other members of its genus, measuring perhaps an inch and a half from the ends of its long antennae to the tip of its tail. It's an impressive insect to find in a moth trap, though not an uncommon one. All *Enicospilus* are nocturnal, and they are frequently attracted to light. When I pinged Natural History Museum taxonomist Gavin Broad—a world expert on Hymenoptera and a keen moth trapper himself—for an identification, he came back not just with the name, but with behavioral information too: "Destroyer of Eggars!"

E. inflexus is a koinobiont and *solitary*—just a single wasp emerges from each Eggar host. The Eggar is a large insect but so too is the wasp, and each needs a whole hairy Eggar caterpillar to develop properly. The wasp is also apparently a specialist. Aside from one record from a Drinker Moth, *inflexus* develops exclusively in Oak Eggars.

There is a very famous, if perhaps apocryphal, quote by the eminent biologist J. B. S. Haldane, responding to the question of what science

---

v. Fairyflies are not flies.

Enicospilus inflexus, identified by its lack of wing sclerites, kinked wing vein, and narrowed (not fat) head.

could tell us about the Creator: "God has an inordinate fondness for beetles."[vi]

It's true that more beetles are known to science than species from any other animal order—more than 350,000 have been described, or about as many as there are known species of plant. However, beetles almost certainly hold their top spot in the list of creation under false pretenses. Taxonomists also have an inordinate fondness for beetles. Collecting these insects was a popular pastime in the nineteenth century, and this has probably disproportionately inflated the number of their entries in catalogues of life. It's likely that wasps would usurp this number-one spot by a substantial margin if they had been as well studied.

The reason we think this is that the Oak Eggar is not alone in having

---

vi. While the veracity of this quote is unknown, Haldane utilizes it in his 1949 book *What is Life?*: "The Creator would appear as endowed with a passion for stars, on the one hand, and for beetles on the other."

its own species of parasitoid. Studies of some well-known insect groups, including close relatives of the Eggar from North America, suggest that there is probably, on average, around one specialist parasitoid for every other species of insect on the planet. In other words, perhaps half of all insect species are parasitoid wasps. Parasitoids even attack other parasitoids. "Only" around 92,500 species of parasitoid Hymenoptera are currently known to science, but the true number is likely to be ten times this figure. In terms of absolute diversity, they are probably the most poorly recorded group of insects on the planet.

Wasps have already pushed beetles from their number-one spot in the UK, where insect diversity is relatively well known overall. Here, the work of taxonomists like Gavin Broad has led to the identification of more than 6,500 species of parasitoid Hymenoptera, compared to just over 4,000 species of beetle (and around 2,500 species of moth). New species are being found in Britain every year—just the number of *Enicospilus* species recorded here has increased by 50 percent this century. Compare that 6,500 to the nine UK species of social Vespid— like the yellowjackets and hornets—that we typically picture when we hear wasps mentioned.

What we should probably conclude about the Creator is that They have an inordinate fondness for parasitoid Hymenoptera. (And given how parasitoids make their living, what should we conclude about the Creator's character?)[vii] It wouldn't surprise me greatly if you hadn't even heard of them before today.

Parasitoids drive the evolution of moth defenses just as much as do birds and bats. Some moths employ physical barriers to prevent attack. The Brown-tip covers its eggs in irritating hairs, and its caterpillars grow up under tents of silk. Some caterpillars can drop away from parasitoids down silken ropes, or wriggle through holes from one side of a leaf to the other. Toxic chemicals sequestered in moth bodies may defend against parasitoids as much as against other predators. If a parasitoid does get through these defenses, all is still not lost. Moths

---

vii. Charles Darwin himself wrote: "I cannot persuade myself that a beneficent and omnipotent God would have designedly created the Ichneumonidæ with the express intention of their feeding within the living bodies of Caterpillars."

can form a hard shell of blood cells and melanin around koinobiont eggs, encapsulating and killing them. However, the huge diversity of moth parasitoids testifies that they are adept at storming moth defenses.

As noted in the preceding chapter, the presence of the right food is not in itself a guarantee that an animal that eats it will be present in an area, but the reverse is more likely to be true. An *inflexus* in the trap is surely a sign of Eggars in the vicinity. It may not have rivaled quantum electrodynamics, but a successful prediction is satisfying in any context.

In ecology, the process of consumption of one animal by another is termed *predation*. The consumer is the predator, and the consumed is the prey. Technically, predation can be considered the flow of energy from prey to predator, and so while it may result in the death of the prey, that is not necessarily the case. Parasites can be viewed as predators in that energy flows to them from their host, though not always with fatal consequences (indeed, the death of a host can often mean death to the parasite, too). Parasitoids are certainly predators under the technical definition.

While competition is a subtle and often cryptic driver of death, predation usually is not. The effects of a predator on its prey are generally obvious, and far from subtle. And moths have many predators. The importance of predation is clearly written on their bodies at all life stages, through their varied evolutionary responses to its pressure. How these effects on individual animals feed through into the dynamics of prey—and predator—populations can move back into the realms of subtlety, though.

To understand the dynamics of predator and prey populations, we typically start by returning to the mathematics of Alfred Lotka and Vito Volterra. Their models helped us to understand possible outcomes when two species are pitted against each other in competition. They also provide the foundation for how ecologists understand the outcome when the interaction is predation.

For competition, we considered what would happen to the population of a species, constrained by limited resources, if another species also depended on those resources. For predation, the resource for one species *is* the population of another. The potential consequences for the death rate of the prey species should be obvious.

The Lotka-Volterra predator–prey models begin with the basic equation for exponential growth that I belabored in chapter 1. They assume that, left to its own devices, the prey population would grow, and without control. Births add to the population, and deaths subtract, but the former outnumber the latter, and the trajectory of the population is ever upwards. Or it would be, if it weren't for the population of the predator. It's the predator that applies the brakes to population growth in the prey, through the extra deaths it causes.

The growth of the prey population, according to Lotka and Volterra, is then the normal exponential increase, but minus the extra deaths— the individuals harvested by the predator. How many individuals are depredated depends on three things—the attack rate of the predator, how many predators there are, and also how many prey there are. The number of prey matters because how likely a predator is to encounter its prey depends on prey abundance—the more prey there are, the easier they will be to find. The number of predators matters, obviously, while the attack rate is a measure of the impact each predator has in depressing the prey population. Individuals of some predator species kill more prey than others, all else being equal.

Assumptions about the consumer are the opposite of the consumed. Lotka and Volterra assumed that the population of predators will be in exponential *decline*, unless predators can find food. Death is a given for them, but only with food can they also reproduce. Without births, extinction is the inevitable outcome. How well the birth side of the population equation goes for the predator depends on how many prey there are, how many predators there are, and how efficient the predator is at turning prey into more predators. If there are more predators, more prey to catch, and more offspring produced for each prey caught, then there will be more predator births in total. Hence, the faster the predator population grows.

So, how large will the population of prey end up, when a predator is adding to its death rate? And how large a population of the predator can the prey support? The answers to both these questions are: It depends. Why is obvious, on reflection.

Predator populations grow by harvesting prey, and how fast they can grow depends on how many prey there are to harvest. But every prey animal consumed means a lower prey population, and that it's harder for predators to find their next meal. The more predators there are, the more they deplete their food supply. Eventually, there are too few prey to support the existing population of predators, and predator deaths begin to outnumber predator births. The predator population begins to decline.

At this point, there are still plenty of predators around, though. They will continue to consume prey—just not quickly enough for their birth rate to offset their death rate. Predator numbers are in decline, but for the time being so too are prey numbers. The prey population is already too low to support the predator population, but is pushed lower still. Predator population numbers follow them down.

Eventually, both predator and prey numbers will drop to low-enough levels that the prey population begins to rebound. When there are few predators and few prey, the chances that they will run into each other is low. Now not enough prey are being killed by predators to control prey population growth. Prey death rate drops below prey birth rate, with the consequences we now know well. At the same time, the predators are not catching enough prey to be able to add more in offspring than they are losing through death. Predator numbers are still in decline. But the population of their prey grows.

As prey population growth starts to soar, predators begin to feel the benefit. They start to encounter more prey, and their birth rate starts to recover. Predator numbers start to increase. At first, while they *are* on the up, they are not yet at a level where they are catching enough prey to offset prey birth rate. Both prey and predator numbers increase. Finally, predator numbers reach the point where their kills *do* outnumber excess prey births. There are still lots of prey, and lots of predators, and so they continue to run into each other, to the detriment of one, and the benefit of the other. Predator numbers carry on up. Numbers of prey, though, are now forced into decline.

Every prey animal consumed means a lower prey population. Every prey animal consumed means that it's harder for predators to find their next prey . . . but we've been here before. We're going round in circles— and if you plot a graph of predator numbers versus prey numbers, so too does the line on that graph.[viii] Sometimes predator and prey numbers both go up (the "southeast" quadrant of the circle), sometimes they both go down (the "northwest"), and sometimes one goes up while the other goes down (the remaining two quadrants). But they return to the starting point and go round again. So says the Lotka-Volterra model of the population dynamics of predator and prey.

Another way to picture the outcome of this model is to think about how predator populations and prey populations vary over time. Both populations show regular fluctuations. Predation is density-dependent, because it depends on the abundance of the prey. The prey population increases until the predator population forces them down. The predator population increases until prey becomes too scarce, when it is starved into decline. Then the prey population can rebound, and the cycle begins again. The predator cycle follows that of its prey, a quarter turn out of step.

Population cycles are a possible outcome of predator–prey models, but not the only one. It's possible that the population of predators is just the right size to cull individuals from a population of prey of just the right size that prey births and predator deaths cancel each other out. Populations of both species would then sit in perfect balance. As I say, it's possible.

Balance is a more likely outcome when the prey has a carrying capacity at which its population would stabilize if the predators were absent. Recall that the carrying capacity is the level at which intraspecific competition causes deaths and births to cancel out. When predator numbers are low, the upward trajectory of the prey population can be slowed instead by competition among prey. This prevents a large overshoot in the prey population, and as a consequence, the predator population also does not overshoot when conditions are right for it to rebound. Population cycles eventually dampen out, and predator and prey populations both settle down to stable levels. The extra deaths imposed by the predator result in a lower prey population size than if the prey were given free rein, though.

The opposite extreme can occur when population fluctuations

---

viii. Well, technically it describes an ellipse, but the effect is the same.

are large—predator populations grow so high that they push prey populations so low that one or both species cannot recover. Population extinction is the result.

Population cycles in both actors are a classic outcome of Lotka and Volterra's predator–prey model. They are one of the main reasons why the models are so interesting to ecologists.

Anyone who runs a moth trap will appreciate that population numbers go up and down. Most species seem to have good years and bad. I caught five times as many Lunar Underwings in 2019 as 2020, despite trapping the same sites over the same part of the species' autumn flight period. When studied over long periods, populations fluctuate from year to year in a manner that looks very much like cycles to the naked eye.

The Oak Eggar is one example. Data from the Rothamsted Insect Survey show that the number of Eggars caught varies substantially from year to year. Catches appear to peak (and trough) every eight years or so. In 2019, I didn't catch any Oak Eggars at all. In 2021, loads. Lotka and Volterra provide one explanation for this—predation. Could good and bad years for Eggars be an effect of their predators—population cycles driven by their specialist destroyer, *Enicospilus inflexus*? Perhaps. But fluctuations are not necessarily cycles—and anyway, we lack data on the parasitoid population, making it impossible to say in this case.

We certainly *do* see population cycles in predators and their prey in the real world, though. The textbook example is Canadian Lynx and Snowshoe Hare.

Lynx were a valuable commodity in the fur trade, and records of the numbers of pelts traded in Canada through the Hudson Bay Company (HBC) identify ebbs and flows in the size of the Lynx population there over two centuries. In years when the Lynx was abundant, tens of thousands of pelts passed through the HBC books, courtesy of a network of trappers. In other years the numbers could drop as low as a few hundred. Statistical analysis of HBC records shows a cycle in the abundance of Canadian Lynx, with an average of ten years between successive peaks (and troughs). Equivalent data for the Snowshoe Hare, key prey for the Lynx, show a similar cycle, with peak abundance a couple of years prior to that of its predator. This is the classic pattern expected from Lotka and Volterra models. Not only do predator and prey populations cycle, but predator peaks follow roughly a quarter cycle later.

Few populations in the real world show such clear and regular population cycles as the Lynx and Hare, but some of those that do are moths. In North America, major outbreaks of Gypsy Moths similarly occur every ten or eleven years, like the Lynx and Hare with which they cohabit. Apple Fruit Moths in Norway cycle more rapidly, peaking every two or three years. Larch Budmoth populations in the Swiss Alps show peaks in abundance around once every nine years—a cycle that may have been turning for a least a thousand years, if analysis of variation in the width of tree rings is to be believed (as with oaks and Winter Moths, the trees do not grow so well in years when the Budmoth is abundant). Ecologists have suggested that predation may be driving all of these patterns.

Positing that predators drive population cycles in their prey is one thing. Proving it is quite another matter. Demonstrating what causes fluctuations in wild animal populations is incredibly difficult and time-consuming, and so there are hardly any species for which we even have a hint of the answer. The moths mentioned above are some of the species for which we do.

One of the best-studied is the Gypsy Moth in North America. The testimonies I reproduced earlier showed the capacity for extreme growth in this population, but it did not—and does not—maintain such plague-like levels of abundance. Most of the time the moth is present in forests in low numbers, but every decade or so its population irrupts, or surges. We can detect regular cycles (as opposed to random fluctuations) by plotting the relationship between the abundance in a year and in all previous years, and seeing if there is any evidence that populations are more similar in size over certain time gaps. For the Gypsy, abundances around nine to eleven years apart show the strongest positive relationships—lots of moths in a given year is likely to signal lots of moths ten years later, give or take. Gypsy Moth populations cycle.

This decadal cycle is accompanied by especially sharp drops in abundance two years after the population peak. A drop with this time lag is what we would expect to see if the prey population were being regulated by a specialist consumer, like a parasitoid wasp. Years of high moth abundance result in lots of opportunities for parasitoids, which are then abundant the following year. The impact of all these extra parasitoids is expressed the year after that, when lots of moths fail to appear—the sort of lagged effect on prey numbers that the Lotka-Volterra model leads us to expect. So the Gypsy moth shows evidence of population cycles driven by a predator.

Well, yes and no.

Patterns in the Gypsy moth population *are* what we would expect a parasitoid to produce, but the problem is simply that we don't see much evidence that they suffer from parasitoids. Rates of parasitism in the population are low. Moreover, there isn't strong evidence that what parasitism we do see is density-dependent—that is, that high moth populations lead to high parasitoid populations. If parasitism isn't density-dependent, then it can't be driving ebb and flow in the host. Ditto if there aren't many parasitoids.

So if parasitoids aren't killing American Gypsy Moths, then what is? The answer turns out to be predators of a different type—mice. Small mammals are major consumers of Gypsy Moth pupae, and this is probably their major cause of death, at least when it isn't an outbreak year for the moth. Scientists have shown that boosting small mammal populations leads to more predation on Gypsy Moths, while trapping those mammals reduces it. So Gypsy Moth population cycles are driven by predation, after all.

Again, yes and no.

I've mentioned before that all models have assumptions, that set of more or less credible beliefs we accept to be true for the purposes of the maths. Lotka and Volterra's predator–prey models include the assumptions of the exponential model underlying them—populations are closed, all individuals are identical, etc.—but with additional ones on top. A key assumption of the predators here is that they are specialists. The predator species only eats the prey species, and unable to catch that prey, it starves. This assumption will be true for quite a lot of predators, including all those specialist parasitoids such as *E. inflexus*—and as we have seen, there are potentially hundreds of thousands of specialists in this group alone. But small mammals are not specialists. If mice can't find Gypsy Moth pupae, they eat something else instead. In fact, they *prefer* to eat other things. When mice can easily find alternative food, predation pressure on Gypsy Moths is lower.

Mice may not be specialists on Gypsy Moths, but they could still drive the Gypsy's population cycles if their consumption of moths were density-dependent. If mice were more likely to eat moths when moths were abundant, that would serve to push high moth populations down. If they were less likely to eat moths when the moths were rare, that would allow the moth populations to bounce back up. Unfortunately, there isn't much evidence that mice consume moths in this way. Mice *are* a major consumer of Gypsy Moths, and do account for a lot of the variation in moth population growth. But this variation depends on how abundant the *mice* are, not the moth.

And finally, this gives us the clue as to how mice actually influence population cycles in their prey. Mouse populations fluctuate, too. When they are abundant, they eat a lot of moths. Moths get their reprieve when mice are rare. The question is, why might mice sometimes be rare? It's not down to Gypsy Moths, but instead, to the availability of other mouse food. The answer is in the trees.

The North American forests invaded by the Gypsy Moths are dominated by oaks. Mighty oaks from little acorns do grow, but not just oaks. Acorns are an important food source for many small mammals, including those that also eat Gypsy Moths. Acorns especially matter in autumn. Mice must eat enough of them to accumulate the fat reserves needed to survive the cold months of winter. Unfortunately for the mice, in some years the acorn crop fails, and when that happens, many mice can't bulk up enough to make it through the winter. Mouse populations crash. This is good news for the Gypsy Moth population. It takes a couple of years, but eventually the lack of mice results in an irruption of Gypsy Moths. These are the years when millions of acres of forest can be defoliated.

So Gypsy Moth population cycles *are* driven by a predator, just in a manner quite different from the mechanism encoded in Lotka and Volterra's population models. Crashes in the acorn crop drive crashes in mice. When the mice crash, moth populations are free to take off. The moths may reach plague proportions, stripping the trees of their foliage.[ix] These moth outbreaks are short-lived, though, because they trigger a plague of their own—the nuclear polyhedrosis virus that we met in a previous chapter. This pathogen rapidly brings the Gypsy population back to low levels. By this point, the mice have recovered enough to keep the diminished moth population in check again. At least until the next time the acorn crop fails.

One last question we need to ask is: Why does the acorn crop fail? The explanation again involves cycles.

Oak trees naturally have their own cycles in seed production. The acorn crop peaks every two to ten years, depending on species. As there are different species of oak in the North American forests invaded by Gypsy Moths, with different timings, and different abundances in different parts of the forest, there are acorns of one sort or another most years. However, from time to time, years of low productivity become synchronized—no oak species produces acorns, and this critical food crop fails. The most likely reason for synchronization is bad weather, which hits all the oaks simultaneously, and over large areas. Exactly why bad weather should occur every ten years or so is unknown. Cycles in sunspot activity have been suggested—these last eleven years or so—and correspond with highs and lows in solar activity. Low solar activity giving a relatively cold year every decade or so might account for the failure of the acorn crop with the same periodicity. It's plausible, if unproven.

As an aside, acorn production is not only of consequence for the Gypsy Moths. People are impacted, too. Good acorn crops in an area boost local mouse populations, but also attract White-tailed Deer to feed on the bounty. Deer are a key food source for Black-legged Ticks, and populations of these parasites grow in turn. Predators and prey again. Unfortunately for us, the ticks act as vectors for Lyme Disease, while the mice act as reservoirs for the bacteria that cause it. Cycles in the acorn crop in northeastern North America not only influence cycles in the numbers of Gypsy Moths, but also in case rates of Lyme disease in people. All of nature is linked.

---

ix. And may stress the oaks enough that the acorn crop fails again in the following year.

As we've revisited the Gypsy Moth, let's return to the Winter Moth as well to see the effects of predators. The Winter Moth also demonstrates how incredibly time-consuming it is to identify the causes of fluctuations in wild animal populations.

The Winter Moth is one of the species that keeps enthusiasts running traps through the long, cold nights of the northern winter, or at least the male does. The female does not come to light, being flightless—indeed, wingless. In late autumn, females hatch from pupae in the soil and climb up to the tree canopy to lay their eggs. These hatch in April so that the caterpillars can catch leaf burst, feed, and pass through their larval instars. Full-grown caterpillars then abseil to the ground to burrow into the soil and pupate. Come autumn, this cycle starts again.

The Winter Moth population in Oxford University's Wytham Woods was studied throughout most of the 1950s and '60s by research teams led by two British entomologists, George Varley and George Gradwell. They set traps on the trunks of five oak trees to catch females on their ascent, so that they could calculate the adult population size. They dissected some of their catch to assess their egg loads, and so estimate how many eggs would be laid by the population. They then set more traps to catch the caterpillars as they dropped, and dissected them to discover how many had parasitoids. How their estimates of numbers changed across all these life stages allowed them to census the Winter Moth population, to identify where death happened, and to infer the major causes. They repeated this painstaking effort for nineteen years. Time-consuming indeed!

All of this work allowed Varley and Gradwell to track changes in the Winter Moth population (at least on their five trees). It turns out that it fluctuates a lot from year to year. In good years, the abseiling caterpillars can be a hundred times as abundant as in bad. Good and bad years are just as much a feature of this species as the Gypsy Moth, albeit with less spectacular outbreaks in the good years.

The Georges' hard work also allowed them to identify the key points in the life cycle for Winter Moth deaths.

Most deaths happen at the first larval stage. The weather is the most likely cause, either by speeding the development of the moths, or delaying the unfurling of the oak leaves. In either case, the young caterpillars emerge to find their timing out, and no food waiting for them. Varley and Gradwell apparently also found evidence that early deaths may be related to the abundance of egg-eating beetles in the genus *Dromius*. If so, Winter Moth losses early in the year could be more down to how predation hits the eggs than to how bad weather hits the newly-hatched caterpillars. Unfortunately, Varley and Gradwell never published their beetle results, so we can't be sure.

Either way, predators definitely matter at the pupal stage. Three-quarters of the Winter Moths entering the soil to pupate never reappeared as adults. Generalist predators (like the mice that eat the Gypsy Moth pupae) and a pupal parasitoid, *Cratichneumon culex*, were the culprits. These deaths were density-dependent—more Winter Moths died under trees with higher pupal populations, and more in years when mortality at the first larval stage was lower. In other words, the pupal predators help to dampen out Winter Moth population fluctuations—they claim more moths in the years when the weather claims fewer (or possibly egg predators do so). This helps to keep the prey population at or close to a stable level—an equilibrium—although the weather adds an overarching element of environmental stochasticity, and thus good years and bad.

On a good night—in summer, warm, and in good habitat—I can pick more than 300 moths out of the trap. It's a lot of work to identify them all, but tessellated they would barely cover the trap's base. Even a "full" trap is far from inundated.

Much of nature is edible, yet most of it goes uneaten. We see our land (by and large) as a vista of green, of plants. Not (by and large) as a writhing mass of caterpillars. Moths are consumers, but most of them barely get the chance before they are themselves consumed. Eggars, Gypsys, Footmen: their populations have innate potential to explode, but that potential mostly goes unfulfilled. Many of the wild animals that we see when we survey the landscape grow fat off the back of this failure. Insects are one of the world's great natural resources.

Population models of the sort I have introduced in the last three chapters are the first steps towards understanding the contents of the moth trap. They reveal the most basic truth underlying biological diversity—that life has a dizzying power to multiply. They then tell us why that power is rarely expressed—multiplication inevitably leads to interactions, and the outcome of those interactions is often that one or other of the players loses out. Or both. Via competition or predation, their potential is reduced or even snuffed out entirely. Death rates rise and birth rates fall, and population growth is brought into check.

Populations do sometimes take off, though. In the extreme, that may be a sign that something significant has changed—as in Medford, Massachusetts, in 1889.[x] However, even "stable" populations fluctuate in size. Species have good years and bad. Sometimes these fluctuations are regular, and we call them cycles. The moth trap samples this variation, and its catches can help to quantify it. The information it provides in turn helps us to understand why.

Variation in the size of animal populations is the very apotheosis of "it's complicated." Even in the absence of interactions among species, stability is not guaranteed. Time lags inherent to the dynamics of populations in seasonal environments, like most moths in the UK, for example, can set populations fluctuating. Those fluctuations can even be chaotic. That is *before* we start to factor in predictably unpredictable elements, like the weather. Add in other species and the levels of complication increase. Add a predator species and we can expect fluctuations again—now in the populations of both consumer and consumed.

Holding these simple models up against the pictures drawn by nature shows that, as always, they are wrong. Their assumptions brush aside important details. But these models are still useful. Testing to see whether the mechanisms they encapsulate work in the real world shows us not only that the real world is more complicated, but why. Bottom-up influences matter to prey populations as much as top-down: predators

---

x. And one reason why species can do so well when they get taken to parts of the world where they don't naturally belong—like the Gypsy Moth that Léopold Trouvelot carried to Massachusetts—is that they leave their enemies behind.

alone do not drive prey cycles. Indeed, predators can equally dampen those fluctuations down. Many predators are not specialists, and the presence of alternative food sources can prevent their populations from crashing when a focal prey species is rare. Mice do not depend on Gypsy Moths alone. The prey then does not necessarily get the break that Lotka and Volterra would lead us to expect.

Population models have the potential for complexity when they consider just one or two animal species. They can already lead to unexpected outcomes, and that is before we add a third (or fourth or fifth) species. Even simple models make the point that we ought to expect life to be complicated. Our attempts to understand what determines populations of even a single species show that it really is!

However, ecological systems do not comprise just one or two, or even three species. My moth trap can catch more than fifty species in a night. I'm already past 500 species in total after only three years, most of them from just two sites.[xi] The study of populations is the foundation of ecology, but ecologists have to admit that once there is much more than a handful of species involved, a different approach is needed. That is the approach we call *community ecology*. Quantifying and understanding how communities of species are structured are key steps towards understanding the contents of the moth trap.

Before I dive into community ecology, though, there is another topic I need to introduce—*life history*. Species can adopt a wide variety of strategies in their efforts to maximize the difference between the number of births and number of deaths in their populations. The result is that species can live their lives in very different ways. Different strategies help to explain the array of forms displayed by even a relatively uniform group like moths, and give important background to questions around how many species can live together in communities. It is to life history that I turn in the next chapter.

---

xi. And these numbers are very far from exceptional.

Chapter 4

The Codling and the Goat
Live Fast and Die Young, or Linger On?

Or I shall live your epitaph to make,
Or you survive when I in earth am rotten
. . . .

— William Shakespeare

Codling Moth (London).

Goat Moth (Devon).

M y moth trapping was born out of a feeling of disconnection. Family life, London, a busy schedule: I wasn't getting enough time to get out and look at nature. The solution was to get nature—or moths, at least—to come to me. There is one obvious downside to this plan, though. I live in Camden. Not to beat about the bush: Inner London is pretty rubbish for moths.

As I've already explained, moth enthusiasts distinguish between "macro" moths and "micro" moths. As the names imply, macros are the larger species, and micros the smaller ones, but the distinction is informal and not one that reflects the evolutionary history of the group. It's also not one that even accurately reflects size, as some micromoths are larger than most macros, while some macros are no larger than many micros. My first-ever Oak Nycteoline had me reaching in error for the field guide to the micros, but it is in fact a small member of the diverse Noctuid family, and hence a macro.

Most people getting into moth trapping start by focusing on the macros. Because they're larger (on the whole), and their plumage better defined, better known, and better illustrated in field guides, they're easier (albeit still a challenge) to name. Most macros can be identified with the naked eye, or with the aid of nothing more sophisticated than a hand lens. Micros, on the other hand, are frequently so similar that we can only name them with certainty by resorting to dissection—using caustic chemicals to dissolve out their distinctive genitals for microscopic inspection. All of this makes macros much more appealing to people like me—amateurs who like to make lists of the species they've found, but don't have the specialist skills for those species that need, as the field guides put it, *gen. det.*—genital determination. Macros are the gateway drug for moth-ers.

The problem I had in London, though, was that there just weren't enough macros in the trap to satisfy my cravings. Macros *do* live there, and they *do* come to the trap, but numbers and diversity are never that high. I immediately found myself poring over the micros, too. I was hooked into a habit that I've frequently regretted since, as I've strained my eyes over tiny and ultimately indeterminable species.

The species that drew me into the micros was the Codling Moth. It's a gorgeous little animal—actually not that little by the standard of micros, about the size and shape of the edible heart of a sunflower seed. They're mostly colored with delicate, wavy bands of gray and brown camouflage, but the upper-rear corner of the wing bears a dusky ocellus (eyespot)—a dark "pupil" partially ringed by a coppery "iris". Like most micros, its wing scales are easily worn off, but the contrast of the ocellus usually remains obvious. The Codling is one of around 400 moth species in the family Tortricidae that have been recorded in Britain, and is common here. I caught six individuals on my first night of London trapping, and it looked to me like it ought to be distinctive enough to identify. It was, and I was caught, too. Micro moths may be small, but they are still beautiful, and fascinating.

That said, the real excitement of the moth trap, for me at least, is still the macros. They are the gateway moths for a reason. Most are subtle shades of brown, yellow, and green, and about the size of a kidney bean. But there is always the chance of catching the species that deviate from this standard—large and dressed in gaudy colors. The Hawk-moths and Tigers, for example. I dream of opening the trap and find myself staring into the empty eyes of the skull-mark on the back of a Death's-Head Hawk-moth. A species that is not only as large as a mouse, but can squeak like one, too. One day, perhaps. A high point of trapping in Devon during the (first) 2020 coronavirus lockdown was catching a Cream-Spot Tiger. It gets its name from the top view, black wings with variable pale blotches. Pretty enough as it goes. From below, though, it's a different animal altogether—mainly a rich, luminescent red, interspersed with black and yellow, that makes the average cardinal look drab.

A high point, but not *the* high point. That was most definitely the Goat.

I can still vividly remember the excitement I felt on June 10, 2020. I had walked out into the field behind the garden in Devon to retrieve the moth trap I'd left there overnight. I had gated the bullocks out of the bottom half acre for a few weeks by that point, and, relieved from their grazing, the grass had reached knee height. There were invariably moths nestled among its stems, and I always started by pacing a wide circle around the trap to try to spot these hidden gems. I enjoy this extra challenge. I didn't need much skill that morning, though, at least for the animal that was the cause of my excitement. There, sitting quietly on a grass stem, was a Goat Moth.

The Goat is a really special moth. In fact, it's a really special creature, period.

First off, it's one of Britain's largest moths, roughly the size and shape of my thumb. Its upperwings are mainly gray-brown, but with silvery patches, and many darker, partially broken lines running between the leading and trailing edges. It has a buff collar to its thorax. At rest, it looks remarkably like a broken piece of lichen-speckled bark. Brilliant camouflage on the trunk of a tree, albeit less so on a grass stem in a field. Wherever it sits, though, it's a stunningly beautiful insect.

It's also a rare species in Britain. It used to be found widely, if patchily, across England, but has notably declined here over the last few decades. Where I trap in southwest Devon, looking across the River Tamar to Cornwall, is one of its last strongholds in the country. Even where it hangs on, the Goat is not a species that often comes to light. Its appearance by the trap was as unexpected as it was exciting.

The Goat is also remarkable for its lifestyle. The female can lay up to about 500 eggs, which she does in batches of fifty or so on the bark of various tree species, including willow, birch, and ash. They especially like trees growing in damp areas, which may explain why southwest Devon remains a stronghold. The caterpillars then crawl under the bark, eventually burrowing into the heartwood, where they feed. When fully grown, they can reach around four inches in length, but they don't rush to get there. The Goat is the slowest of all British moths, passing three or four winters inside its tree before pupating. The adult finally emerges in the summer, four or five years into its life.

The species is actually named for the "goatish" smell of its leisurely caterpillars, but for me, the Goat will always be an acronym: Greatest Of All Time. Big, beautiful, rare, and exceptional is a compelling combination in any animal.

A species like the Goat Moth raises some interesting questions. Contrast it with the Codling. Adult Codling Moths emerge in late spring to mate, and in the case of the females, to lay her few dozen eggs. Codling caterpillars also burrow into their food, but in this case a fruit— usually an apple, but sometimes a pear, a quince, or something else. They grow quickly, and within about a month are fully grown at a less impressive (but still, from the perspective of fruit growers, annoyingly large) inch or so. At this point they leave the apple and head to the soil to lie dormant for the winter. They pupate the following spring. Sometimes they can squeeze two generations into a year in Britain. In the more amenable climes of the Middle East, they can manage four or five. Unlike the Goat, they do not dawdle.

Two species, both clearly stamped from the classic moth template. Not dissimilar in shape and color, at least. But two species with very different ways of *being* moths, or technically, with very different *life histories*. One large, long-lived, and fecund, but rare. The other small, producing fewer offspring as it cycles through its generations quickly, but far more common. Most moths lie somewhere in between these two ways of life—though the Codling's end of the spectrum is more crowded with species. But why? Why isn't every moth more like a Goat?

The answers emerge from the processes that we have seen in play in the previous chapters. Birth and death. Consumption and being consumed. How these processes vary and interact mean that there is more than one solution to the problem of being a moth. A pleasing diversity of form in the moth trap is the result.

We know the potential to grow that is inherent in all animal populations. We've seen what that potential can do when given its head, from the repopulation of Britain by Footmen, to the irruption of the Gypsy Moth in Massachusetts. And that's just moths. The lice and locusts that gave us Biblical plagues still plague us today. But it could be so, so much worse.

Evolutionary biologists have an imaginary ideal organism—they call it a *Darwinian Demon*. The aim of every organism is to leave as many copies of its genes in future generations of its species as it can—to maximize its Darwinian fitness (a concept we met in chapter 2). The Darwinian Demon would be the result if a species could perfect every aspect of reproductive performance in achieving that aim. It could start reproducing from the moment of its birth. Its offspring would be legion, and each one as well-endowed as its parent in its capability for instant and copious reproduction. It would continue into immortality. The Demon is a model of exponential population growth, but untroubled by death, and with a birth rate unparalleled in nature.

The Darwinian Demon is a thought experiment. It's something we imagine, to help us understand why it only exists in imagination, thankfully. For the Demon, there are several important reasons why. Even before we consider all the myriad bringers of death to an animal, the real world imposes constraints that make the perfect impossible.

All of life runs on energy—the fuel that it gets from its diet. Plants and other autotrophs (with a few notable exceptions) harvest solar energy. Other species rely on the hydrocarbons that autotrophs manufacture using that solar energy—consuming autotrophs directly, or consuming the consumers—and run their metabolic systems essentially by burning these pre-fossil fuels. The amount of energy that any animal can acquire is limited,[i] however, because consumption takes time—to locate, take in, and digest food—and also requires some of the energy acquired previously. All animals are therefore running off a finite supply. Just as with household income, they must work within a budget. They have to make choices about how to expend the precious fruits of their consumption.

The decisions that animals make in framing their life histories are guided by the ultimate goal of those choices—leaving as many copies of genes as possible in future generations.[ii] All animal populations must produce at least as many offspring as they have deaths if they are not to decline to extinction. For a stable population of most sexually reproducing species, the minimum requirement is to raise two offspring per female, on average, to reproductive adulthood. As we've seen, all species have the capacity to raise more when conditions are in their favor, but any population that persists is doing at least the minimum. Given that no species is Demonic, how they go about replacing themselves with offspring is what requires choice.

Why not just produce as many offspring as possible, as quickly as possible—that is, do the best impression of the Darwinian Demon? This is certainly the strategy adopted by some species. Moving closer to our ancestors, the House Mouse is a good example. A female Mouse is sexually mature at about six weeks of age. Her pregnancy lasts around three weeks, and she can give birth to up to fourteen pups, each weighing less than a tenth of an ounce. Three weeks later, and they're weaned. A female can easily do this five times a year. And when conditions are favorable, mouse populations can reach plague proportions. There is a photograph of a pile of mice caught in one night at Pullut in the Australian state of Victoria, labelled "200,000—3½ tons." House Mice are alien in Australia—like the Gypsy Moth in America—and plagues happen in the wheatbelt there every three years or so. (There's a big one happening as I type this.) Female House Mice generally don't live more than a year in the wild, though, even when stressed farmers

---

i. As are other essential resources.

ii. These decisions are not conscious (despite the phrasing), but are driven by natural selection in the struggle for existence.

are not making concerted efforts to kill them. They live fast and die young.

The House Mouse makes one set of decisions about allocating energy to reproduction, and it's hard to argue with its success, but there's more than one way to make a living. At the opposite end of the spectrum is the African Elephant. It takes a female Elephant around ten to twelve years to reach sexual maturity, and then will generally give birth to a single calf, weighing around 220 pounds, after a pregnancy lasting almost two years. The calf can be weaned after six to eighteen months, but it may continue to suckle for several years. Females typically only give birth every three to six years, but can live to be seventy, and reproduce until late in their life. Elephants live slow and hope they don't die before they get old. Yet while they may not rush to reproduce, their populations are just as capable as House Mice of growing exponentially when conditions are good. Indeed, I use the African Elephant population in Kruger National Park as an example in my lectures.

Mice and elephants lie at opposite ends of the mammal spectrum in terms of how they go about allocating energy to ensure as many copies of their genes as possible in future generations. They reinforce the lesson of the Codling and the Goat—that there is no single route to success. They also highlight what species must choose between when they make their decisions: survival, growth, and reproduction. No species can have them all. That inevitably leads to trade-offs. Which way a species goes depends on the ever-present specter of death.

Death comes to us all. Most of us don't know the appointed hour. However, while we might not be able to predict exactly when we will die, we do know the broad distribution of probabilities. That's the same for other species—though through the blind influence of natural selection, rather than the conscious interpretation of mortality data. When death comes matters.

Although animals can die at any time, we can broadly distinguish between death occurring in two phases of life: juvenile and adult. Juvenile mortality is death that comes before an animal is capable of reproducing, while adult mortality is death that comes afterwards. The distribution of death across the different stages of life has important consequences for how animals evolve to live. Ecologists have thought about this in most depth for mammals, perhaps because of our awareness of our own mortality.

Let's think about a female mammal with a certain reserve of energy to invest in producing offspring. In an ideal world she would like to produce lots of well-endowed babies, all with good prospects of surviving to produce babies of their own (and grandchildren for her). Unfortunately, that option is not available to her, because her reserves are limited. The choices narrow: produce lots of offspring with less expenditure on each, or spend more on each of fewer offspring (or give everything just to one). This is *the* classic life history trade-off that animals (or at least mammals and birds)[iii] must make: few large offspring versus many small. Quantity versus quality. Which end of the spectrum the mother chooses has consequences. More offspring means smaller offspring. They can mature more quickly, but each has a lower chance of making it to adulthood. Investing more into each means larger offspring that each have a better chance of survival, but they need longer in the womb, and take longer to mature once out of it.

Here's the rub. It's only sensible to make a longer-term investment if you're going to see that investment mature. A mammal needs its mother around, at least in its early days, if it's going to survive. Definitely before it's born. If a mother's survival prospects are poor, then it's a big risk to take the slow route to a few well-endowed babies. The result may be reproductive failure. When death is likely, a better strategy is to produce as many offspring as she can, as quickly as she can. Make hay while the sun shines.

Going all out in reproduction is not always the best strategy, though, or all mammals would act more or less like mice. There are definite advantages to a slower pace of life.

For a start, better-endowed offspring tend to be superior competitors. We've already seen that, in a finite world, it's likely that some will lose out in the fight for food or other scarce resources. It pays to be

---

iii. More on this soon.

better equipped to compete, whether with your kin, or with individuals of a different species. Quality is also likely to matter when it comes to competition for mates. Parents that are likely to live to see their grandchildren will give their own offspring as good a start in life as possible.

A slower pace of life may actually increase the chances of living to see those grandchildren. Reproduction is exhausting, and going all out may lower the chances that the mother will survive. It makes sense to keep something in reserve in order to help evade predators or fight off disease, for example. Or to help make it through lean times—winter or drought, say—to get to the next breeding season. The trade-off here is between devoting energy to reproduction or energy to survival. The choice may also be between energy to reproduction or energy to growth. Growth may be a better bet in the long term. When they finish growing, larger adults can invest more in their offspring. Larger adults may be able to fight off predators to which they would succumb if less robust, and thereby up their survival chances. Planning for future births may increase the probability of living to see them.

Not rushing to start reproducing may be advantageous, too. There are all sorts of reasons why more-mature females make better mothers. Extra time spent in growth and development can deliver females that produce larger offspring, survive better, and go on to be more productive parents themselves. As long as the benefits of delaying reproduction outweigh the costs—mainly the risk of dying before getting the chance—then a more leisurely approach to parenthood is likely to be the right choice.

As mammals, we see these influences on our own lives and life choices. For much of the evolutionary history of humanity, life has been nasty, brutish, and short. That has changed, though, with better, more-reliable food supplies and the birth of modern medicine. In the last sixty years, life expectancy worldwide has increased from fifty-three to seventy-three years of age, and infant mortality has dropped from sixty-five to twenty-eight deaths per thousand births. As the chances of dying young, and before reproducing, have decreased, so too has the birth rate. The average number of births per woman halved from 5 to 2.5 in the same period. Women are having fewer babies, and having their first baby later. People have been putting more into growth, too, with the average woman born in 1996 three inches taller than those born in 1896 (the difference is four inches for men). As mortality rates have dropped, *Homo sapiens* has moved towards a slower pace of life. Just like other mammals, our growth, reproduction, and survival are all linked. We still follow nature's rules.

Growth, reproduction, and survival are all linked in moths, too, but moths are not mammals. We still expect moths to suffer from the same ultimate constraints as do we, and indeed all other organisms. They have limited energy budgets, and they must make decisions about how to spend them in order to maximize their Darwinian fitness—copies of genes in future generations. However, moths grow and reproduce very differently from us, and death takes them in different ways. This does not mean that moths are following different theoretical pathways through life to mammals. Just that the balance of factors is different, and the end points different, too.

Moths and mammals are similar in some features of how they live their lives. Larger moths lay larger eggs, which hatch into larger caterpillars. Larger caterpillars take longer to develop and emerge from their pupae as larger adults. Those adults themselves have longer lifespans. Moths line up along a continuum of life histories, from small and fast to large and slow—from Codling to Goat—just like mammals.

Where moths and mammals diverge is in how reproduction lines up with their pace of life. For mammals, slow means large offspring, but few. For moths, slow also means large offspring, and more of them. The classic reproductive trade-off—few small versus many large—is not one that moths apparently have to make. Imagine if elephants produced litters of dozens of calves. That is essentially what Elephant Hawk-moths do.

Why are moths like mammals, except when they are not? We are not sure yet, but it looks like differences in parenting skills are likely to play a role. So, too, are differences in how different approaches to life affect the likelihood of death.

Let's start with parenting.

Moths actually don't do much of it, and that's probably important. They provision their eggs, and a larger egg can cram in more material. That will benefit the caterpillar that develops within it—just as a longer gestation period does for neonate mammals. A well-provisioned egg is especially important for species that overwinter within its shell, and they lay larger eggs than otherwise would be expected. Other than filling their eggs, though, most moth parents do little beyond finding a good place to lay them. Some don't even really do that, instead scattering their eggs in roughly the right place as they fly over, like an old-fashioned bomber. Moths in the family Hepialid—Swifts and Ghosts, for example—use this approach.

If you're not doing much parenting, how many offspring you have is not constrained by your ability to look after them. The advantage of being large is now not that you can protect and rear a small number of larger offspring, but that you can produce and sow a large number of larger offspring. It's win–win, as far as reproduction goes.

The importance of parenting in life history choices can be seen in those insects that aren't as carefree as moths. Some do show parental care, at least in terms of providing food for their developing larvae. Examples include the dung beetles, which roll up balls of dung for their larvae to feed on, and carrion beetles, which find and bury carcasses for their young. Carrion beetles frequently turn up in the moth trap, although their arrival is not always greeted with pleasure. They are notoriously foul-smelling and frequently crawling with phoretic (hitchhiking) mites.[iv] Carrion beetles can be caring parents, though. They will even digest flesh and regurgitate the resulting meaty soup to begging larvae, in a manner similar to how some birds feed their chicks. This is relevant here because insects like these, which do care for their offspring, also show a trade-off between offspring size and number, just like mammals. Parenting, or the lack of it, matters for the life history strategies that animals can adopt.

If being large, but without childcare responsibility, allows you to win in the number of offspring you can produce *and* win in the size of those offspring, there's clearly a benefit to being large. This explains why we

---

iv. Just the thought of them makes my head itch.

see Goat Moths. But most moths that arrive in the trap don't have the impressive bulk of the Goat—indeed, it's very much the reverse. Small is more the order of the night. Why then are most moths like the Codling, and few like the Goat?

The answer comes back to death.

As we've already seen, there's a lot that eats moths. Birds, bats, and beetles. Tens of thousands of parasitoid wasp species. Even mice get in on the act. Then there are the diseases, like nuclear polyhedrosis virus (NPV). And it's not just the species that eat them that kill moths—competition does for them too. Moths have evolved all sorts of countermeasures against predators and ways to avoid competition, but the bottom line is that most moth eggs don't survive to lay eggs of their own.

In the circumstances, it's a dangerous strategy for a moth to dawdle through life. When there are so many other animals out to get you, you need to crank through your life cycle as quickly as you can. For juveniles, that means growing to adulthood as fast as possible. Since bulking up takes time, most moths have to forego that option in favor of making it to maturity. Smaller adults may not be able to produce so many eggs, but at least they can produce *some* eggs. Those small adults themselves have shorter lifespans, and so need to lay their eggs quickly. At least they don't have so many to lay.

For mammals, there is cause and effect in both directions in the relationship between death and size. Elephants gain protection from predators through their strategy of growing to large size. They can then invest time and effort in raising the next generation. We expect to see strategies like this evolve when juvenile mortality is high relative to adult mortality—adults have the time to invest, and using it improves the life chances of their offspring. Unfortunately for moths, size is no protection from death. Even the largest are easy prey for most of their bird and mammal predators. It's true that larger moths are probably better able to escape the clutches of most spiders—battering their way through their webs—but that's of small consolation if they are then picked off by nightjars or bats. In fact, large moths are likely to be *more* at risk from vertebrate predators—they're easier to spot, and more valuable prey. Being small is probably a better strategy to avoid being eaten, if you're a moth. You're harder to spot, and barely worth the effort of looking for. That's probably why, accounting for other differences, day-flying moths are smaller than night-flying moths.

Size is also no protection against parasitoids or disease. The Oak Eggar is a large moth, but its large caterpillars simply fall victim to that large parasitoid, *Enicospilus inflexus*. NPV particles sit on caterpillar food, waiting to be consumed. If anything, bigger consumers will be more at risk.

There's another problem with growing to be large if you're a moth: food supply. It's often ephemeral, and that also doesn't encourage the taking of time. The Codling Moth is a good example. Its caterpillars burrow into fruit in late summer. They're not appreciated by orchard farmers, causing what's known as "maggoty apple." If you've ever bitten into a Cox to find a brownish tunnel and half a caterpillar, then it may well be a Codling Moth you've just depredated.[v] The apple or other fruit is the Codling caterpillar's world for the duration—or at least until they've eaten their fill. But apples are not a long-term proposition as a home. They're meant to have a short shelf life before (ideally) being consumed, and their seeds dispersed, by a frugivore. The Codling caterpillar must get in and out in quick time so as not to be consumed along with the fruit. This constrains how long it can grow, and how large it can get in the time available. The same problems are faced with those making a meal of apple leaves, like the Winter Moth (which doesn't only eat oak leaves), or the Apple Leaf Miner (which, as you will have guessed, has the additional constraint of needing to live between the surfaces of a leaf).

The whole plant can in many cases be as ephemeral as the leaves or fruit of an apple. Remember all those moths I catch in London that feed on a "a wide range of herbaceous plants?" Those herbaceous plants—the Docks and Nettles and Plantains—are themselves at the fast end of the spectrum of life history strategies. They have to rush through growth and reproduction and move on quickly before shrubs and trees take over. The herbs are quick to colonize and exploit empty space but are poor competitors—they will lose that space once those larger, slower plants

---

v. Google "Codling Moth," and the page is topped by ads for traps of very different designs and intent to the one I run on the roof terrace.

arrive. Herbaceous plants are a common and widespread food source for insects, but one that requires an equally rapid turnaround from any herbivore that wants to exploit them. It's no surprise that, all else being equal, moths that feed on herbaceous plants are smaller than those that feed on trees. Again, the caterpillars must get in and out in quick time, and faster means smaller.

When the food supply is long-lasting, though, moth options widen. And as far as food goes, there isn't much that lasts longer than wood. Indeed, durability is pretty much the point of it. Trees can live for decades, centuries even, on the skeleton that wood provides them. A moth with caterpillars that feed on wood can take its time growing to maturity, safe in the knowledge that its food supply is unlikely to disappear. Indeed, they may *have* to take their time, because while wood is reliable food, it's hard to make a living on. Lots of species can consume and digest soft leaf or fruit matter. Fruits are even *designed* to be eaten. Relatively few species have the processing capabilities required to chew through tough cellulose and turn it into animal tissue. Even for those that do, it can take ages to extract enough nutrition to grow—especially to grow large. One of those that has mastered this way of life, of course, is the Goat Moth—and it still takes this moth five years to complete its long, slow life cycle.

As well as being reliable, wood also provides species living inside it with protection from other causes of death. Bad weather is unlikely to be a problem for a Goat Moth caterpillar, unless it's bad enough to kill the tree. Casual predation is also off the menu—it takes specialist work to locate and consume a Goat caterpillar, though there are of course parasitoids that can do it. Ichneumonid wasps like *Stenarella gladiator* and *Lissonota setosa* come equipped with long, sharp ovipositors to drill through wood and into the caterpillars.[vi] Larvae of the parasitoid fly *Xylotachina diluta* crawl down the tunnels made by the Goat Moth caterpillars to find their prey. No life is free of risk. Despite these jeopardies, wood provides the Goat Moth with food and protection, and allows it the time to invest in growth. Its impressive size is the result.

---

vi. An ovipositor is the delivery tube through which a female insect deposits her egg.

That large abdomen has room for more eggs, and the adult moths live long enough to lay them. Size *and* number.

All species must decide what to do with the time they have. Constraints imposed by the need to grow, reproduce, and survive all interact to determine the course they plot. There is no one right answer to the question of how to be a moth, or indeed any other organism. Different circumstances lead to different solutions. The different answers contribute to producing the diversity of forms that come to a moth trap.

The different answers have other consequences, as well. An important one is how species can respond to uncertainty and the vagaries of chance.

We've already met one form of this in environmental stochasticity—that random element of variation in the weather, for example, or in habitat availability, that has the potential to take even healthy, growing populations down to levels where extinction is a real possibility. Or even beyond them. The original Gypsy Moth colonists in Medford were lucky not to get hit by bad weather in the years when their numbers were still low. Even now, when they've spread widely across northeastern North America, bad weather has substantial effects on their populations—albeit positive ones—through its impacts on the acorn crop, and the populations of mice that would otherwise keep the Gypsy Moth in check.

But stochasticity is not just environmental. It can also be *demographic*. This is randomness that arises because the fundamental processes of birth and death are also inherently chancy. Just as the concept of the "normal" environment glosses over the existence of important variation, so too do the concepts of "the" birth and "the" death rate mask the fact that chance plays a role here, too. An unlucky run of unexpected deaths, births that by chance are mainly males—demographic stochasticity can also bring a population close to extinction. It's a particular problem when populations are small, when every birth and death really matters. Large populations can generally take a bit of bad luck in their stride.

Escaping from the perils of demographic stochasticity may be one of the advantages of having a fast life history. Populations of fast species can generally grow quickly at low densities. Their numbers can build rapidly, especially when they can run through several generations in a year. The less time a population spends at low numbers, the less likely it is that demographic stochasticity will come to bite it. The flip side is that demographic stochasticity is obviously likely to be a concern for species with slow life histories, if their populations fall to low numbers.

What knocks populations down to low numbers, though? One of the main causes is the bad times we get through environmental stochasticity. Populations of fast species can grow rapidly, but they can also quickly get knocked down again—if their shorter lifespans coincide with a period of poor weather, for example. Species with slow life histories, on the other hand, play a longer game. Their populations can ride out short-term variation in the environment, and so are less likely to be knocked down when times are bad. A spring storm can strip apple trees of their blossoms, and come autumn there will be no apples for Codling Moth caterpillars. A cold spring can find a forest-full of Winter Moth caterpillars hatched out but without oak leaves to eat. And as we have seen, their population crashes. But a poor spring will not bother the Goat Moth caterpillars sitting deep within the wood of the same trees. Slow species may be more susceptible to demographic stochasticity at low population sizes, but they are less likely to be knocked down to those levels by chance.

Living slow is one way to get through environmental variation, by riding out the bad times. Another is to hedge your bets—to back more than one horse in the race. Species can invest in a mixed portfolio of life history strategies on the basis that different approaches pay off under different circumstances. A classic example of bet hedging is plants that produce seeds with long dormant periods. The strategy of a Darwinian Demon would be to develop to reproduce as quickly as possible, but sometimes it pays to wait. Paying into a bank of seeds ensures that a plant has savings to bolster against unpredictable bad times in the future.

Moths also don't have to put all their eggs in the same basket. Indeed, this is one way that they can hedge their bets. Females can choose to lay eggs in one large clutch (or a few), or many smaller clutches. The female Vaporer Moth, for example, is flightless, and lays all of her hundreds of eggs on the cocoon from which she eclosed. The spectacular ermine-clad Puss Moth, in contrast, lays her eggs in batches of two or three (they are destined to hatch into equally spectacular caterpillars). Which choice a species makes probably depends in part on the likelihood that a clutch will be destroyed—many small clutches are better if most clutches don't make it. Females can spread their risk around their habitat.

Moths can also spread risk by not relying on just one sort of food—a single species of plant, for example. Remember all those species whose caterpillars eat a wide range of herbaceous plants? More options reduce the likelihood that caterpillars will starve in the event that the moth emerges into a bad patch or year for any one foodstuff. A broader diet may also hedge against the nasty, brutish, and short lives of the parents. Short-lived adults tend to have more-catholic caterpillars, which makes it easier for the mother to find suitable plants for her offspring in what little time she has.

Species can also hedge their bets over time, as with the seedbanks of plants. Many species of moths appear in the trap over many weeks or even months, even though adult lifespans are much shorter than this. Take the Autumnal Moth, for example, a species of high latitudes in Europe and North America. Adults live for about two weeks, but the species' flight period is around three months. The moth spreads out its appearance across autumn because different individuals pass different amounts of time in the pupa—some a month, others as many as three. Autumn weather is predictably unpredictable in the latitudes inhabited by the Autumnal Moth. Its strategy of having offspring with different pupal periods reduces the risk that all of a female's offspring will appear during a bad spell.

The strategy of hedging bets over time is taken to its extreme by *Prodoxus y-inversus,* one of many species of Yucca moth. Yuccas are well-known, among other things, for having obligate pollination relationships with moths. No Yucca moths, no Yucca seeds. *Prodoxus y-inversus* is not one of the pollinators, but its life history is no less remarkable. Fully developed caterpillars of this species undergo a period of dormancy in Yucca fruits, which normally lasts from summer until the following spring. This in itself is not unusual—many moths have dormant periods in order to ride out the winter or other times of shortage. The Codling is one example. However, if *P. y-inversus* caterpillars don't get a period of cold weather over winter, their dormancy doesn't get broken, and they stay as mature caterpillars for more than one year. In fact, they can remain dormant for up to ten years if conditions don't give them the chilling they need. Take their seed homes inside, and the moths can stay dormant for up to thirty years. But chill them after all this time, and off they go again. The desert homes of the Yucca and its moth can go through long periods when key resources, like water, are scarce. *P. y-inversus* deals with that uncertainty by banking offspring in the same way that plants bank seeds—indeed, it banks them *in* seeds.

A final way in which species can hedge their bets is by spreading themselves widely across the environment. Some individuals stay local, but others take to the wing and head off to look for riches elsewhere. This is a strategy with significant consequences for moth populations and communities, which I'll explore at length in a later chapter.

London may be pretty rubbish for moths, but set out a trap on my Camden roof terrace and a wide variety of forms will still appear. It's true that most of them are middling-to-small and fifty shades of brown—very much the middle of the road of moth construction—but even in urban London some buck this trend. I'm not holding my breath for a Goat, but Lime Hawk-moths are stunning creatures to find on an inner-city roof terrace, and while common, Oak Beauty, Yellow Shell, Chocolate Tip, Rosy Footman, and Angle Shades all speak of different ways of being a moth. The diversity of life is mind-bogglingly immense, presenting a daunting task to even begin to understand. Paring back the question to just one part of this complexity is a major undertaking. Yet I think—I hope—that the first few pixels in the grand picture are starting to fall into place. Some semblance of order is starting to appear out of the darkness.

So far, we have really only been working with three premises: organisms are born, organisms die, and those births and deaths happen on a planet of varying but finite resources. Yet these simple truths underpin all of ecology and evolution. They form the basis of all that follows.

Finite resources interact with deaths and births to determine how many individuals there are in a population. But they also determine what those individuals *do* between birth and death. To live fast and die young, or to live slow and get old—these are the ends of the life history continuum on which all species must find their place. Codling Moth towards the fast end, Goat towards the slow. Where species lie depends on the opportunities that death gives them to consume, and how they channel their consumption to growth and reproduction. The choices that species make contribute to the diversity of forms in the moth trap—Codling Moth and Goat, and all the wonderful variety between and beyond. These choices explain why biologists are as obsessed with size as journalists are with age.

Birth, death, and finite resources have given us a good start in making sense of the diversity of moths that come to a moth trap, giving us a grounding in the ebb and flow of populations and why species live their lives differently. Now, however, we need to move on. We need to address how and why we see collections of certain species in the moth trap. Why some species and not others. Why we find the numbers of species we do.

We need to consider the ecology of communities.

Chapter 5

The Uncertain
Living Together in Communities

The rich man in his castle,
The poor man at his gate,
God made them, high or lowly,
And ordered their estate.
 — Cecil Frances Alexander

Vine's Rustic (left) and Uncertain (I think), Camden, London.

discovered the joy of trapping moths relatively late in life, but I've always been a biophile. Apparently, even before I could talk, I would stand by the window of our suburban house and point at the birds flying by. For as long as I can remember, I've known that whatever direction life took me, I wanted birds to be involved in some way. Hard work, combined with fat slices of luck, got me to a job where I could study them for a living. And the travel that is (or until recently was) a fortunate bonus of academia meant that I've been able to go birding in all sorts of stunning locations around the world. Being out in the wilds, mind emptied of everything except the sounds, sights, and smells of nature, is meditation and therapy combined.

The first year of the Covid-19 pandemic taught us many things, but one of the most positive from the perspective of an ecologist was a better appreciation of the natural world. Many of us had our movements restricted by Covid-19 regulations, and so our houses and (if we were lucky enough to have them) gardens and parks, for long periods, became our whole world. The huge drop in human traffic allowed nature a temporary opportunity to reclaim the spaces that we would normally monopolize, and this did not go unnoticed. People saw and heard the wildlife emboldened by our retreat to our homes, and they realized the beauty inherent to instinctive and unaffected creatures. Nature provided beams of light in a dark year. People connected with it.

Trapping moths is one way we can go further in forging the connection—not just passively observing, but actively bringing nature to us. It's a hobby that burgeoned in Britain as an antidote to the lockdowns we suffered here. According to Dr. Zoë Randle of the charity Butterfly Conservation, 2020 saw a 62 percent increase in records submitted to the Devon Moth Group, and a 72 percent increase in the number of people submitting those records.[1] That's a lot of new people taking an interest in moths. I can testify to the joy that catching these fragile and fleeting creatures can bring.

We share our world with an incredible array of species—from viruses to whales—but not all groups are equally attractive to the amateur naturalist. The most popular is birds. Large numbers of people take more than a passing interest in them; the main British bird conservation charity, the Royal Society for the Protection of Birds (RSPB), has more members than the three main British political parties combined. More than half of all homeowners here put out food for garden birds. There are thousands of twitchers—competitive birdwatchers—who will (pandemic regulations permitting) drive, boat, or fly hundreds of miles to see rare birds that have gotten lost and appeared accidentally in Britain (and this is not just a British affliction). I do it myself from time to time, I admit. Birds may be the most popular, but other groups high up on the league table of interest include butterflies, dragonflies, bees, mammals, plants, and of course, moths.

Why do these groups get the most attention from wildlife watchers? It's actually a balance of features. They need to be visible, for a start—to attract attention, and to make them amenable to identification and recording (most amateur naturalists have a strong instinct to make lists). Ideally, finding them should require little in the way of specialist training or equipment—a pair of binoculars, for example, is both cheap and versatile. There needs to be enough diversity to hold people's attention, but not *too* much. Too few species, and one can quickly find and familiarize oneself with most or all of them (the six native British species of reptile and eight amphibians present little challenge to the keen lister); too many and the task of chasing them all down becomes overwhelming (amateur parasitoid wasp enthusiasts are few). Finally, one needs to be able to identify the species, again without much specialist equipment or training. Ideally, there's a field guide. Ideally, species should be distinguishable. But here's the rub—ideally, there should also be degrees of difficulty. Easy species to suck you in. Harder ones to test your developing skills. And puzzlers to present a real challenge.

If it were too easy, it wouldn't be fun.

Birds really hit the sweet spot for these criteria, and this is one reason for their preeminence among wildlife watchers. But so too do moths—or the macros, at least. Many birders are moth-ers as well. Macromoths are easy enough to see, with the use (and one-off cost) of a light trap and a hand lens. Moth-ing is popular around the world, but Britain is especially fertile ground for the hobby. This country is home to around 900 species—enough for years of recording, but it's not an impossible dream to see them all. They can be found anywhere, so a roof terrace in London will get you going. But some species are restricted to certain habitats or locations—an ideal incentive to visit new places and take your moth trap on holiday. There's a great field guide—more than one, in fact, with gorgeous paintings and photographs of every British species, plus text outlining when and where to find them. Some are easy to identify. No one is going to have trouble with their first Buff-tip. But others—they can be a challenge indeed.

Some of these challenges have gone so far as to have been recognized in the names of the moths themselves. Britain has the Suspected. It has the Confused (by us—I'm sure it gets on by itself just fine). And it has a species that I was faced with in my first-ever London moth trap—the Uncertain. Or at least, I'm pretty sure that's what it was.

British lepidopterists have long been challenged by the Uncertain. It's an archetypal moth from the diverse family Noctuidae, about the size of a thumbnail. At rest, its forewings fold across each other like the jacket of a double-breasted suit. It's typically a light-brown color, with slightly darker brown, white-rimmed oval and kidney marks on each forewing, and three slightly darker brown crosslines running from the leading to the trailing edges.

The problem is, this description could equally apply to a species very closely related to the Uncertain, called the Rustic. Its forewings look a little smoother and shinier. The crosslines are a bit less distinct. The kidney mark and oval are closer in color to the brown background of the forewing on which they sit. But all that is comparing pristine specimens. A little wear and tear can roughen forewings and obscure the contrast of crosslines and kidneys. In such cases identification becomes, well, uncertain. At least without resorting to *gen. det.* Even fresh individuals are hard to do by eye. I logged on to Twitter and asked @MothIDUK to help me with my first.

If there were just a pair of confusing species for the Uncertain it would be bad enough. Recently, though, a third has entered the mix—Vine's Rustic.

Vine's Rustic is very close to the Rustic and Uncertain in pelage. Compared to the other two, its overall color is grayer, its appearance chalkier, its oval and kidney marks a little larger, and the leading edge of its forewing straighter. The differences are all rather subtle, though.

Vine's wasn't an identification problem for British moth-ers until the start of the twentieth century, when it began colonizing southern England. Once it established a foothold it spread rapidly, and it now occurs widely south of the line connecting the Bristol Channel on the west coast, and the Wash on the east. It outnumbers the Uncertain in my trap in London, although the reverse is true in Devon. That I often catch them together at least helps me tell them apart.

The Uncertain, Rustic, and Vine's Rustic pose challenges of identification, but they pose another challenge, too—one that relates to coexistence.

I catch all three species in my trap in Devon, while two of the three appear commonly on my roof terrace in London (not yet the Rustic). They all look essentially the same—I doubt that any predator could distinguish them on color or pattern. They all appear to have essentially the same lifestyle. All are moths of a wide range of lowland habitats. All three are out feeding as caterpillars from midsummer through to autumn. All feed on that by now familiar salad of a "range of herbaceous plants," and their diets overlap: the field guide lists chickweeds, docks, dead-nettles, and primrose for the Uncertain; chickweeds, docks, and plantains for the Rustic; and docks, dandelion, prickly lettuce, and primrose for Vine's. All overwinter as caterpillars before pupating in spring in an underground cocoon. The adults are all on the wing at the same time of year.[i]

How is it that three such similar species of moth—all doing essentially the same thing—can live side by side in the same area?[ii] This is a fundamental conundrum for ecologists. It's the question of what determines the composition of ecological communities.

---

i. However, Vine's tends to appear earlier in the year, and also flies later into the autumn.

ii. If you're really keen, they can be distinguished by dissecting their genitalia— their distinctive internal anatomies make the species sexually incompatible, and thus definitely different!

Understanding communities—how nature structures them—is a difficult question, even by ecological standards. What even *is* "an ecological community"?

On the face of it, that's not too difficult a question, although there are several definitions that we could use. The most general is that a community is the set of populations of living organisms in a given area over a given period of time. This recalls the definition of a population, but with groups of species replacing groups of individuals. We have moved up a level of ecological complexity.

No ecologist considers *all* populations in an area, though. The task would simply be too great, except for the very smallest of plots. We tend to focus on populations from a specific taxon—plants say, or birds. Or moths. Trying to understand the community structure for any one of these groups is generally considered a difficult enough proposition—and it does anyway inevitably draw in considerations of other groups through their interactions with our species of interest. Predators of moths will be relevant, for example. So too moth food. Some definitions of the community explicitly invoke interactions, as (you may again recall) do some definitions of ecology as a whole.

As with populations, ecologists delineate communities in space and time, and also as with populations, largely for our own convenience. Areas should be neither too small to represent meaningful sets of organisms, nor so large that they encompass multiple sets of interacting organisms—multiple communities. This already hints at problems ahead, because if by definition a community is the set of populations in a given area, then how *can* the given area include multiple communities? Are communities just collections of organisms that happen to be living together, or is there some deeper, more fundamental integrity to them? This question goes to the heart of debates about how communities are structured. More of that anon.

Similar issues pertain to time. We need to study communities over periods relevant to the species of interest, but what *is* a relevant period? As with populations, communities are dynamic entities, but changing membership is more problematic. The species present in an area naturally change over time—as with the arrival in England of Vine's Rustic—but how much change do we allow before we have a different community? It's an ecological Ship of Theseus.

The difficulties faced by ecologists crystallize when we move on to consider the key features of the community that we are trying to understand. There are many aspects that we could explore, but we can view three fundamental and overarching questions as core. How many species can live together? Which species are they? And how common are the different species in the community—more formally, what are their *relative abundances*? We can use the information provided by the species in my moth trap to think about these questions a little more.

Let's start with the first question—how many species live together in the moth community around my London roof terrace? Well, I've caught 245 species there all told, macros and micros. So that's the number we need to explain, right?

Well, probably not.

I've caught 245 species, but it's unlikely I've caught every species in the community. I've only been trapping for a couple of years. What I have is a *sample* of the community. How complete that sample is I don't know. Here's the first problem—how do we begin to understand how many species there are in a community if we don't know how many species there are in a community?!

The best we can do is to *estimate* the number. Whole books have been written on how to do that, so you probably won't be surprised to discover that there's no single or easy answer. We generally go about it by looking to see how the number of species trapped has increased over time. Usually when we sample a community, we catch a lot of species quite quickly, but eventually the law of diminishing returns kicks in and we find new species less and less often. We can use the accumulation of species over time (or more accurately, over increasing numbers of samples) to predict the number at which we'll stop getting new ones at all. This is then our best guess at the species richness of the community. Applying a couple of different statistical methods for doing this to my

London trap predicts that I'll run out of new moths once I hit around 340 or 350 species. Doing the same for the garden I sample in Devon produces estimates of around 390 to 410 species.

It makes sense to compare estimates of species numbers, rather than the actual numbers themselves. I've spent less time trapping in Devon than London (139 nights versus 204) but caught more species of moths there—326 to date. I'd expect my efforts to plateau at a higher number in Devon, and that's what the statistics predict, too. Unfortunately, we never know what the true number is. We can keep on trapping, essentially forever, but that's when the time issue starts to rear its ugly head.

Communities change over time. We expect this, for all sorts of reasons. Remember chapter 2—pollution killed the lichens in cities, and the Footmen disappeared. Clean air legislation has resulted in both returning. Thinking longer term, most of Britain was covered with ice just 12,000 years ago. The moth community on the banks of the Thames will have altered a lot since then. Trap "the" moth community of Camden Borough long enough and it will not really be "a" community at all.

As it goes, the precise number of species in Camden (or Devon, or wherever) doesn't really matter. We always expect some error when we sample the universe (even physicists do). What matters is the relative numbers, the general patterns. What we have to explain is why we get fewer moth species in London than Devon. Of course, we need to compare more than two sites to come up with robust answers to questions like that.

Moving on, what about the relative abundances of the species in the moth community around my London roof terrace? Or to put it another way, how are the individuals I catch distributed among the species I catch? Are there common and rare species, or are all species more or less as abundant as each other? Once again, we have to think about this question while remembering that my moth trap shows me a sample of the community, not the community itself.

The genesis of the study of how individuals are distributed among species—*species abundance distributions*, as they are known—owes much to moths. Specifically, to a light trap run by the entomologist C. B. Williams at Rothamsted Experimental Research Station in Harpenden, the small town in southern England where by coincidence I spent my teenage years. C.B., as he was known to his friends, was interested in the statistical patterns displayed in biodiversity. The abundances of moths in his trap must have drawn his attention like, well, a moth to a flame. Also employed at Rothamsted at the time was the brilliant statistician (but unrepentant eugenicist) Ronald Fisher. It was Fisher to whom C.B. took his moth trap records. Their joint work on species abundances (co-authored also with lepidopterist Steven Corbett from the British Museum) is arguably the foundational work on species abundance distributions.[iii]

C.B.'s moths showed two key features. First, the rarest abundance class—those species for which only one moth had been trapped—was the most frequent. More moths were represented by one individual than any other number. Second, there were successively fewer species with two, three, four, etc., individuals caught. Plot a graph of number of species versus number of individuals and join the dots, and you get a smoothly curving decline away from the peak at one individual (like a J pushed onto its back). Based on this pattern, Fisher described a statistical relationship for the expected number of species with different numbers of individuals, known as the log-series. In a log-series distribution, the numbers of species with two, three, four, etc., individuals are regularly smaller proportions of the number of species with singletons—with one individual.

The log-series distribution describes—and implies—regularity in how individuals are divided up among species. Most species are very rare, and few are very common. It was an exciting development in the history of understanding patterns in biodiversity. Yet it implies a fact

---

iii. It's not the first work, though. An earlier paper on the topic by Isao Motomura was overlooked. English is the *lingua franca* of science, and Motomura's work was published in his native Japanese. At least Motomura's work is now part of the discussion, although the model of the species abundance distribution that he proposed is (like Fisher's log-series) not believed to be a good descriptor of the underlying reality.

that many dispute—that the most frequent sort of species is the very rarest.

The abundance most often recorded in C.B.'s moths was one individual. The same is true for my moth trap, in both London and Devon. Yet of course there will be more than one moth of most of these species—the moth trap reveals a sample of the community, not the community itself. If we continue to trap—and this indeed happened for C.B.'s moths—we will continue to catch more individuals of some of those singleton species. We will also catch individuals of species that were too rare to be caught given the original sampling effort—species that, as we discussed above, we know to be missing from our species lists. Eventually, if we caught everything, we would find that there are some very common species, a greater number that are very rare, but that most species sit somewhere in between. Most species are uncommon (but most individuals belong to the common species). In cases where we know essentially the whole community—British birds, for example—this is what we see.

Larger and larger samples of a community "unveil" the shape of the species abundance distribution. They get closer and closer to the true form. Unfortunately, we never actually see the true form—even our British bird abundances are estimates, albeit extremely good ones. This allows ecologists to expend a lot of effort arguing about what that true form may be, largely based on assumptions about unseen individuals from unseen species. It also allows them to champion a variety of different models for how individuals divide up among the species— models describing different mechanisms determining abundances in the real world. The truth here is also veiled.

For now, though, it's enough to know the bottom line, which is that some species are common, some species are rare, and most species are somewhere in between. *Exactly* where varies a lot from community to community.

So community ecologists are trying to explain the number of species when they don't know how many species there are, and species abundances when they don't know how many species *or* individuals there are. What about the third key question—which species can live together in communities? Do I need to point out that this is also a difficult question to answer when we don't know all the species? It's relatively easy to estimate the *number* of missing species, but much harder to say *which* species are missing. On the plus side, we do know that the missing species are likely to be rare. At least we won't be missing much.

Right, I think I may have belabored the point—ecological communities are naturally difficult to understand. That's enough excuses. What *does* determine their composition?

As you might be expecting, it's complicated.

Let's start by going back to the last time we thought about similar species sharing the environment—interspecific competition. Lotka and Volterra's models showed that pairs of species could win, lose, or draw in competition. It's the circumstances of the draw that matter from the point of view of species being able to live together in communities. Remember that competitors coexist when intraspecific competition matters more to them than does interspecific competition. Species have to find ways to live where their opposition is weak, so that the greater struggle is with their own kind. This is true for pairs of competing populations. But it's also true more widely. When a species has a way of living where it's better at doing its own thing than any competitor, it will be able to persist in a community with those competitors.

Coexistence needs competition between members of the same species to be stronger than competition between members of different species for two reasons. On the one hand, stronger intraspecific competition allows a species to recover if its population drops low for any reason. Bad weather, say. If interspecific competition were stronger, then it wouldn't get the chance to bounce back—the competitor would simply press home its newfound advantage. On the other hand, stronger intraspecific competition means that a species controls its own population. It stops one species pushing out others when its population gets high. Stronger intraspecific competition *stabilizes* communities.

Consumers can stabilize communities, too, when the community includes their food. Specialist predators can hold down populations of species that would otherwise be strong competitors. We exploit this in biocontrol—species like the cactus moth *Cactoblastis cactorum* introduced to Australia to chew down enormous infestations of Prickly Pear that would otherwise outcompete native plants. Generalist predators, like the mice we met in chapter 3, can have the same effect if they mainly eat common prey species. This prevents any one species from getting too common, but allows species respite if their populations get knocked down to low levels.

One way to think about a species' way of living is that this is its niche. We met this concept earlier, when I defined it as the set of conditions that a species requires to persist, but it has made its way into common parlance as "the ecological role of an organism in a community, especially in regard to food consumption." This more ordered view of the niche dovetails neatly with the idea that a species needs to be best at doing something to coexist with others. We can imagine a community, then, as a set of species, each with its own distinct way of life. This always brings to my mind a Georgian village idyll, with shopkeeper, publican, doctor, farrier—each with a preordained role that is theirs alone. Everyone knows their place in the ordered estate, a picture of deterministic order. This *niche-structured* model of communities is the classic view.

Did such villages ever exist? I doubt it. There must always have been tensions bubbling under the picturesque veneer. The idea of an ecological community as an orderly set of tessellated niches also glosses over a messier reality. The niche as commonly understood is really the *realized* niche—the set of resources a species can use when it is interacting with other species. What we observe a species doing is less preordained and more pragmatic—it occupies the position it could get. For the Uncertain, the realized niche includes a wide range of food plants and habitats in which it can complete its life cycle, in a wide range of environments spanning much of Eurasia from Europe across to Mongolia and Korea. This is what the species does do, given all the competitors and predators it must contend with. Admittedly, that's quite a lot. What the species *could* do, though, if given free rein, we don't know. What it could do—the set of resources it could exploit in the absence of competition or other negative interactions—is its *fundamental* niche.

Many species—most?—have the potential to do much more in a community than we see them do. Interactions restrict that potential. We saw in chapter 2 that moths that live between the surfaces of oak leaves are forced into a summer lifestyle by competition with moths that chew up those leaves, like the Winter Moth. Competition can also be *apparent*, with predators or parasitoids killing one species to the benefit of others. As long as niches do not overlap completely—as long as some of the fundamental niche is realized—a species will be able to persist in a community.

How different do niches have to be, then? How much overlap is allowed?

Imagine again two species competing for exactly the same resources— they have the same fundamental niches. If one species is a much better competitor, then it will win the fight, and the worse competitor will go extinct. Realized niche zero. When a species is a poor competitor, it needs to have a niche that is quite different to the competition, or it will quickly be out. This difference matters less, though, when the two species are more or less equally good in competition. They can now both hold their own, even when they are fighting over more or less exactly the same resources. So species can coexist in a community either by doing different things (stabilization), or by being equally good at more or less the same thing—*equalization*. The greater the mismatch in competitive ability between species, the greater the difference in their niches needs to be, if they are to cohabit.

The niche-structured view of ecological communities holds that stabilization is king. Niches must be different for species to coexist. The opposite view may be true, though—that equalization is key. Under this view, codified in *Neutral Theory*, all species are equal—equally good competitors, and all with the same fundamental niche. It sounds like heresy.[iv]

---

iv. And indeed, it was substantially at odds with the established orthodoxy when it was proposed, at the turn of the millennium, by American ecologist Stephen Hubbell, a plant ecologist interested in rainforest trees.

If Neutral Theory is right, communities appear as the result of random processes played out among groups of species that are ecologically equivalent. Species drift in and out of communities by chance, and the appearance of rigid structure is just an illusion. It's ecological anarchy—the antithesis of the Georgian village idyll. But it does get at why one moth trap can turn up three more or less identical species like the Rustic, Vine's Rustic, and Uncertain. Species don't have to be different to coexist—they just have to be equally good.

Neutral Theory assumes that there is some large but fixed number of individuals, of some number of species, that can exist in a region. This is called the *metacommunity*. All the moths in Britain, say. What we think of as communities—the moths around my roof terrace in Camden, for example—are then parts of this whole. When a moth in Camden dies, it creates a vacancy in the community. That vacancy can be filled either by another Camden moth, or by a moth from elsewhere—from the wider metacommunity. Which species fills the gap, and from where, is a chance event, but one that depends on abundance. Common species from the community are more likely to add their offspring to the community. Immigrants are more likely to be from commoner species in the metacommunity. Over time, births, deaths, and immigrants will cause the community to drift randomly to a composition that depends on its size—how many individuals there are—and how connected it is to the metacommunity—how many deaths are replaced by locals versus immigrants.

The composition of the metacommunity is obviously going to matter to all this, because that's where the immigrants come from. This depends on how large the metacommunity is, and how often new species appear by speciation. Speciation matters because the random processes of Neutral Theory cause species to meander to extinction. The advantage of being a common species is that you have more tickets in the lottery to fill an empty space in the community—and so get more common. For most species, the disadvantage of being rare is continually losing out in this lottery, and eventually disappearing altogether. Speciation adds in new species. Without it, we would eventually end up with a metacommunity with just one, superabundant species. Size matters—as usual in ecology—because a metacommunity with more individuals allows more species to coexist.

The great thing about Neutral Theory—and where it really scores over niche structuring—is that it makes predictions about *how* communities should look. About how many species should coexist together, and about their species abundance distributions—the fundamental patterns of community ecology. Exactly how communities are predicted to look depends on the features I've just described. Metacommunities and communities can be different sizes, connected by immigration differently, and speciation can happen at different rates—all of these differences matter.

I'll just give one example. Neutral Theory predicts that the shape of a community's species abundance distribution will depend on the immigration rate from the wider metacommunity. When a community is isolated, the action of chance within it leads to common species dominating, and rare species dropping out. Deaths are replaced by local births, and these tend to come from commoner species in the community. Increasing immigration allows rare species to persist in the community, though, thanks to the individuals arriving from elsewhere. When most spaces appearing in the community are filled by immigrants, rare species become the commonest sort—the species abundance distribution has the log-series shape that Fisher predicted. Immigration rates somewhere in between predict what we actually normally see in real communities—that some species are common, some species are rare, and most species are somewhere in between.[v]

Where Neutral Theory falls down, though, is that species are not all the same. They really *do* differ, and they differ in ways that matter for their ecology. Goat and Codling Moths were not created equal.

Imagine that the community of moths around my Camden roof terrace really was a fixed size. Could the space created by the death of a moth—the Oak Beauty that blundered into the clutches of a spider on my terrace last night, say—*really* be filled by any other species at

---

v. Neutral Theory can also estimate what precise numbers "some" and "most" actually are.

random? It's unlikely. The food plants used by moths matter, and we know that not all caterpillars can consume all plants—despite the many that can consume a wide range of them. Moths are also not all equally good competitors—remember the leaf-mining moths forced into a summer lifestyle by competition with leaf chewers? Identity matters.

Do we also believe that the community of moths around my Camden roof terrace really is a fixed size? That's unlikely, too.

We've already seen that there are good years and bad years for species of moths—think Gypsy—but the same is true for the whole community. My undergraduate lecturers at Manchester University led an annual field course at Woodchester Park in Gloucestershire, and for twenty-four years ran moth traps there.[vi] From 1969 to 1976, they often caught more than a thousand moths per night. From 1977 to 1981, though, they were lucky if they trapped more than a hundred—although numbers picked up again in 1982. They put the variation down to bad weather, which we've also already seen can drive down moth populations, directly or through its knock-on effects on predators. Either way, there's just more year-to-year variation in moth trap data than Neutral Theory can explain.

Competing philosophies of community structure have long been a feature of ecology.

At the turn of the twentieth century, debate centered on the opposing views of two American plant ecologists—Frederic Clements and Henry Gleason. Clements argued that communities are predictable and repeatable units, comprising set groups of mutually dependent species. Gleason's view was that communities are not fixed, but variable, chance associations of species that arise out of how individual species use the environment. We have an illusion of repeatable associations because those individual requirements coincide for some species.

At the turn of the twenty-first century, similar arguments continued,

---

vi. This must have been my first exposure to moth trapping, although I'm sad to say I have no memory of it.

but now framed in terms of niches and neutrality. Neutral Theory emerged from studies of rainforest trees. Here it makes sense that the death of an individual creates a gap in the community—indeed, it literally does—that could be filled by more or less any other tree species. Most communities don't work like that, however.

The truth is likely to lie somewhere between the extremes. Ecological communities are not anarchic collectives of equals assembled by blind chance. Neither are they Georgian village idylls populated with preordained roles. Species have different fundamental niches, but chance plays a role, too—in which species get to coexist, and in how niches get realized.[vii]

Let's imagine a space empty of all the species of interest. Someone has sprayed insecticide over Camden and eliminated all the moths. Over time, they begin to recolonize, and a community assembles. What does it look like?

The first species in gets first pick of the available resources. Its population starts out small but is free to grow. Grow it does, until intraspecific competition kicks in and begins to slow it down. Logistic growth. Eventually, carrying capacity is reached in the niche that the species can realize there.

The second species to arrive does the same thing, as long as there are available resources for it. Likewise, the third, fourth and so on. Each colonizing species needs to be able to grow to escape the perils of demographic stochasticity—the chance effects we met in chapter 4 that can snuff out small populations. If it can do that, then it gets a foothold in the community, and becomes part of it.

But if a species arrives to find its resources already being used—essentially, its niche already occupied—it has a problem. New arrivals inevitably start out rare, and have to be able to grow their populations away from the low levels where stochasticity is a danger. That's difficult

---

vii. Another model—*stochastic niche theory*—proposed by yet another American plant ecologist—Dave Tilman—neatly distills the essence of this interplay. I'm not saying this model is right, but I do think it's useful. Plant ecologists seem to dominate debates about the mechanism determining community structure. This is presumably because plant communities are easier to study—their members don't run or fly away or only come out at night.

if a similar species is already present, monopolizing the resources it needs. Possession is nine-tenths of this particular ecological law.[viii] Common beats rare, through weight of numbers. The colonist cannot get a foothold and fails to join the community.

New species can only join if their requirements are sufficiently different from the incumbents to allow their populations to grow. If its way of living is better than any competitor. Intraspecific competition then stabilizes its membership, preventing it pushing out other species but allowing it to rebound from low population sizes. Specialist predators, like *Enicospilus inflexus* on the Oak Eggar, can also promote coexistence by eating down the populations of species that might otherwise force out competitors.

Obviously, the more species there are in a community, the harder it becomes for new species to squeeze in. There's an element of chance in which species arrive when, and so which species get to join. First come, first served, as it were. Or in ecological parlance, a *priority effect*. Going back to the Rustic and the Uncertain, locations tend to turn up mainly one species, or mainly the other. This could well be a priority effect— which one you find in your moth trap depends on which moved into the neighborhood first.

It's not a completely closed shop, though. From time to time, even well-established communities can be colonized. Vine's Rustic into Camden is an example. Why?

Environmental stochasticity is one likely cause. Bad years can knock moth populations down (or other populations important to moths, like mice), but bad years are not necessarily bad for all species. An unexpectedly low population of a resident species may let a competitor gain that elusive foothold. Especially if it's a species on the up in the surrounding area. Once in, it may hold its own—if the species are equally good competitors, then they should be able to coexist. Equalization. The community has grown by one.

Longer-term changes to the environment will also affect membership of the communities that live there. The impacts of Clean Air Acts on lichens allowed Footmen, and other lichenivorous species, to (re)

---

viii. Disclaimer: it's not a law in the scientific sense.

colonize cities, joining their moth communities. Human impacts on the climate are another way in which we may be influencing membership. The greenhouse gasses our activities emit are warming the planet and increasing the frequency of heat waves. Heat waves are an important element of the environmental stochasticity that can knock resident populations down. Higher average temperatures may then favor colonists with fundamental niches more suited to warmer conditions. I'd be surprised if this were not at least part of the reason for the march of Vine's Rustic across southern England—a species that's historically had more of a southerly bias to its distribution in Europe than the Uncertain.[ix] The future of the Uncertain in the face of these changes may be a case of nominative determinism.

If a species can only join a community if there are resources for it, then we can predict how common it will become—and that depends on how common are the resources that it needs. A moth that feeds on an abundant plant will itself be abundant. Rare species can be perfectly good competitors, just for resources that are themselves rare. The range of abundances that a community displays then reflects how the range of available resources is used. We will find that some species are common, some species are rare, and most species are somewhere in between, when that's also true of resources.[x]

Theories of community structure are many and varied. It would be possible to write whole books on this subject alone—indeed, people have. I think that there are three key messages to take away from all that research.

First, no species can do everything perfectly. As we saw in the last chapter, trade-offs always need to be made—and so each has its own fundamental niche. This defines the conditions under which a population of the species *could* persist.

---

ix. As we will see in a later chapter, the northerly spread of certain moth species across Britain has been a feature of communities in recent decades.

x. That's only a really strong prediction when we know how resources are distributed, though. We usually don't in any detail.

Second, competition is inevitable, but a species *should* persist as long as it has a way of life—a corner of its niche—where it does better than (or at least as well as) any competitor. Where intraspecific competition matters more than interspecific in limiting a species' numbers.

What that way of life is—how much of the fundamental niche gets realized—determines how abundant a species is likely to be.

But third, chance plays a big role, too. Which species get to coexist has an element of luck, and this also knocks-on into how fundamental niches get realized.

These three tenets interact in nature to determine how species coexist in communities. That's the theory, at least. What does it mean in practice, though?

To start with, because niches matter, the more types of resources there are for species to exploit, the more species will coexist.

Yet resources are not necessarily constant. They can vary, in both time and space. This variation affects the type of resources that are available, it affects the diversity of resources, and it affects the abundance of resources. This in turn affects how abundant consumers can be, and the impact of chance upon their populations. All of this has an impact on the features of the community of consumers. To understand how this plays out, we can start on my roof terrace in Camden.

Camden is a seasonal environment. The good times of spring, summer, and autumn are interspersed with the bad of winter. At least this variation is broadly predictable, so that the species that live in Camden will have evolved ways to ride out the winter they know is coming—like the Codling Moth caterpillars that leave their apples and lie dormant in the soil.

Across space, resources in Camden are patchy, at best. There are some large, plant-rich gardens and areas of waste ground (awaiting the developers), and Hampstead Heath is a short walk away. But this is all interspersed with inhospitable roads, paved and concreted patios and, of course, houses.

The interaction of variation in space and time exacerbates patchiness. My terrace gives me eye-level views of the canopies of mature Pear and Cherry trees—a constant in the landscape here for as long as the Victorian houses. It overlooks lawns that are mowed to a low sward with annoying regularity. Aside from grass, little other than moss or Dandelions can get a root-hold on those areas, and gardeners usually do their best to remove even this small diversity. Farther afield, the rangers of the Corporation of the City of London allow some grasses to develop to flower on Hampstead Heath, along with Thistles, Cranesbills, Vetch, and other herbaceous plants of fertile grasslands. They still get their autumn mow, and this regular disturbance means that nothing as substantial as a shrub gets going there. The Heath meadows are interspersed with hedges and clumps of mature trees, however. A patchwork of frequent disturbance, and long-term stability, which affects the diversity, abundance, and identity of resources—and so too of consumers, and the contents of my moth trap.

Some of this patchiness is good. Gardens are home to more moths, and more species of moth, when they have a greater diversity of microhabitats—features like lawns, log piles, ponds, trees, hedges, and different plant species favored as food by caterpillars or adults. The composition of the gardens themselves may actually matter less than the wider area, though—more-diverse surroundings can boost the contents of traps in gardens that would otherwise be unexciting for moths. Most moth caterpillars are herbivores, and as we saw earlier, the species that commonly appear on my roof terrace in Camden reflect the common food plants in the wider area of gardens and parks. If I couldn't see a Horse Chestnut tree from my terrace, I'd know from the leaf-mining moths I catch that there's one nearby. Diverse resources make for diverse consumers. But if you want a full and diverse moth trap, don't just garden for wildlife yourself—rope in your neighbors, too. And maybe chat with your local park keepers.

Patchiness in time can breed diversity, too. Seasons bring distinctive opportunities, allowing species to divide up niches over the year. Recall the Winter Moth caterpillars chewing on the spring flush of Oak leaves, versus the Leaf-miners tunneling between the surfaces of those leaves that made it past the chewers to summer maturity. The moth trap reveals very different sets of species on the wing in Camden in spring versus autumn. When discussing the identification challenges of the Uncertain, I didn't mention the Common Quaker, even though a written description of this species would be pretty much identical to that for Vine's Rustic. Common Quaker is not generally confused with Vine's Rustic—or Rustic or Uncertain—because it's a classic species of spring, and so appears in the trap earlier in the year than those three.[xi] Its caterpillars feed earlier in the year, too. Moth-ers follow the seasons through the species in their traps as well as by the length of the night.

There are flip sides to patchiness, though.

One is that pieces of good habitat may be set in inhospitable surroundings. Clumps of trees in a desert of pesticide-sprayed farmland, for example. Or in a sea of houses. Moth communities are poorer in gardens that include more artificial surfaces, or are surrounded by them—and moth traps in urban gardens yield smaller catches of fewer species. Concrete and brick do not grow moths. Tell me about it. My trap and I only really come alive when we leave the city—on our forays to rural Devon, for example. If you want more moths in your garden—and who wouldn't?—lift those flagstones and let plants take over.

Another flip side to patchiness is that size matters when it comes to habitat. Larger gardens are home to larger and richer moth communities. It's not just how *many* different opportunities they have, but how *much* there is of each. More habitat overall equals more moths. This is one way in which chance plays its role. Small patches of habitat are home to smaller populations, and smaller populations are more likely to fall afoul of stochasticity—environmental or demographic. Species will disappear from these small patches by chance, and may not come back. Their moth communities will be poorer as a consequence.

The patchiness of the London Borough of Camden also influences the types of moths found there. We've explored models for how communities are structured, in terms of number and abundance of species—but have not thought much about *which* species are found in those communities. The same processes matter for identity: niches, competition, and chance, and how they interact with resources.

Small patches of habitat set in an inhospitable matrix of concrete favor species with catholic tastes—*generalists*. Cities are home to moth species with broader diets and habitat preferences than their country

---

xi. It's not unknown for a confused Quaker to appear at the wrong time of year, though—identification should always be based on looks, not date.

cousins. All those species like the Uncertain, whose caterpillars eat a "range of herbaceous plants," for example. Cities are not bereft of *specialists*, but they're fewer in number. When the presence and location of food is uncertain, and resources are thin on the ground, it helps to guarantee a meal if you're not too fussy. The option to develop on a variety of food plants also potentially bolsters a species' population size, working to keep it above the danger levels for stochastic effects. Uncertainty in resources plays against those species with more leisurely lifestyles, exemplified by the Goat we met in the last chapter. City dwelling moths need to hustle, to chase and exploit patches of food while they can. Too slow and the chance may be gone.

Contrast Camden with the site of the best night of moth trapping I've experienced to date: Lewtrenchard Manor, a granite-gray Jacobean mansion on the edge of Dartmoor. Once the home of Sabine Baring Gould—who gave us "Onward Christian Soldiers"—it's now a beautiful country house hotel.[xii] We stay regularly (if infrequently) as a special family treat, as it was the location of our wedding reception. The owners are accommodating indeed, and let me run a moth trap in the woodland on their grounds, including one warm August night in 2020. Lewtrenchard isn't a recognized moth hotspot, but still—what a night that was! The following morning my little actinic was heaving—275 individuals of fifty-eight species. Fourteen were species I'd never caught before—such delights as Scalloped and Barred Hook-tips, Vestal, Small Mottled Willow, and Gothic. Rustic *and* Uncertain in the same trap.

Lewtrenchard differs from Camden in many ways that enrich its moth community.

For a start, the manor sits like a natural rocky outcrop in an extensive tract of woodland. Mature beech and oak trees towered over the moth trap, all nestled in a dense herbaceous understory. Size matters. Moth communities are richer in more-extensive patches of woodland, and richer in woodlands with more plant species. Size matters in the vertical

---

xii. This recommendation is unsolicited. No money has changed hands.

plane, too. Woodlands add structural complexity to landscapes, taking life up into the sky, and allowing each individual tree to harvest solar energy from a greater volume than would be possible if they sat low to the ground. More volume of plant, more food for moths. More moths.

Trees are not just producers of food, but also substrates. Those at Lewtrenchard are green with moss and lichen, alive with opportunities for caterpillars that graze the trunks and branches. It will come as no surprise that I pulled fifteen Footmen out of the trap that August morning, of three different species. The diversity of resources the plants provide permit more opportunities for species to realize part of their niche, and so are home to a wider range of moth consumers. More habitat also means more individuals, and the larger moth populations are less susceptible to the vagaries of chance. Species are better able to persist, and so more can coexist.

Woodland is extensive around Lewtrenchard, but not continuous— it forms a patchwork with fields of pasture and long grass, edged with hedges, brambles, and other shrubs. The moth trap was close enough to a field that I emptied it with an audience of three curious cows. All of these habitats add to the local diversity and structural complexity of producers. They feed moths that leach into the woodland, attracted by the light. In small patches of trees, pasture moths bolster communities lacking in species typical of woodland. This is more the case in Camden. On the edge of forest tracts, though, pasture moths simply add to the richness of the species caught. And unlike in Camden, surfaces around Lewtrenchard incapable of growing moths are few.

Large patches of habitat also affect *which* species are found— especially if that habitat is around for the long-term, as is mature woodland. They allow specialist consumers to take up residence.

The hook-tips are good examples. The Barred Hook-tip is a yellow-brown moth, with a darker central bar across its wings. It rests with its upper wings flat, and these end in the characteristic hook that gives species in this group their common name. It's a species of beech woodland, where its caterpillars feed on, and pupate between, leaves of this tree. Beech is the dominant tree at Lewtrenchard—not least along the footpath heading east from the manor known as Madam's Walk. It's an ideal spot for the Barred Hook-tip. Its Scalloped relative is similar, but with more raggedy trailing edges to its hooked wings, which it tents over its body at rest. Scalloped Hook-tip is also a specialist, but of birch, which is common at Lewtrenchard around the margins of poorly drained meadows. Both beech and birch trees can be found within a few hundred yards of my flat in Camden, but not in great numbers. If they're also home to these Hook-tips, I haven't yet struck lucky. And not through want of trying.

The presence of specialists in a community does not mean that generalists are excluded, and the Lewtrenchard moth community has plenty. The Rustic and Uncertain, for example. Remember that species with similar niches can coexist as long as they are more or less equally good at exploiting their common resources—in this case, a similar range of herbaceous plants at the same time of year. The dense understory carpets the woodland at Lewtrenchard in herbaceous plants, allowing large and persistent populations of even potentially close competitors to coexist. Broad diets and habitat preferences, and the ability to move fast to exploit ephemeral food plants, are no bar to membership of communities in larger and more-stable habitats—they just allow such species to exploit unpromising areas as well. Those that enter the unpromising mothscapes of cities are but a small subset of the communities in the better habitats beyond.

Ecological communities are complicated entities. The number of potential combinations of species living together in any one is far greater than the number of atoms in the universe, even in a relatively species-poor assemblage like British moths. It's no surprise that it's proved difficult to understand how they work. At least now I think we understand the general processes at play.

Communities are structured by the interplay of determinism and chance. On the one hand, we have the concept of the niche—a place for everything, everything in its place. On the other, the luck that puts some individuals (and species) in the right place at the right time, while others miss out. Which matters most—towards which end of the continuum between order and chance communities lie—probably varies from community to community. It will depend on the sorts of species involved, and how vulnerable they are to the vagaries of the environment in which they are embedded.

Whatever, or wherever they are, how species interact matters, too. Competitors coexist when intraspecific competition matters more to them than does interspecific competition. Species need to control their own abundance through intraspecific competition, so that they can recover from low numbers but not become so abundant that they overwhelm other species. We met this sort of density-dependent control in chapter 1. Predators can exert it, as well. But even similar species can coexist, as long as they both do the same thing equally well.

Interactions between consumers and their food matter for the consumer communities. More resources allow more species to coexist. Greater resource diversity gives more niches. Greater amounts of those resources help consumers avoid the perils of low population size—and support fussy eaters as well as catholic consumers. Abundant resources can support abundant consumers and explain how individuals are distributed across species in the community.

It's in the specifics where I think this field of ecological science is less satisfying though.

Take the Uncertain, Rustic, and Vine's Rustic. They can probably coexist because they all do slightly different things—favoring a different mix of food plants, for example. They can probably coexist because they are all more or less equally good competitors for those resources. Stabilization and/or equalization. We probably tend to get mainly Uncertain or mainly Rustic in any one trap for reasons of chance. First come, first served, perhaps. It's unsatisfying that we don't really know.

That *is* typical of ecology more widely, of course. We know for a few species that competition matters—Cinnabar Moth and Ragwort Seed-head Fly, for example—but the huge amount of work needed to demonstrate competitive effects conclusively means we're never going to know how—or even if—most species interact. We know that predator populations can drive variation in numbers of their prey, but interactions among predators, environment, and prey are so complex that we can only guess at the relative influence of top-down and bottom-up effects for most species. It's no wonder that we can only guess at the specifics of how the Uncertain, Rustic, and Vine's Rustic coexist. I should be happy that we know the generalities.

And I am. But I want to know more. I want to know why the Camden moth community is around 350 species, and why Devon is around 400. Not just why we expect to see more species in Devon—although at least having a general idea for that is good. Likewise, I don't just want to know that some species are common and some (more) are rare (with most somewhere in between), I want to know why the commonest two species in Devon are roughly twice as common as the third, and four times as common as the fourth.

This is why Neutral Theory is so attractive. It can put numbers to community species richness. It can explain patterns in abundance. It's just unfortunate that the way it does it—that all species are equivalent— means it must be wrong. A beautiful theory let down by ugly facts. The fundamental conundrum for ecologists persists.

Neutral Theory may be wrong in detail, but there is one thing it has exactly right. When we define a community as the set of populations of living organisms in a given area over a given period of time, we are tacitly acknowledging that there is a larger area, and a longer period of time. We ignore that wider context for convenience, but really it can't be ignored.

Local communities, like that sampled by my moth trap, are fragments of wider wholes. They may be isolated from other communities, but they are never entirely separate. Individuals can be born into the communities we define for our convenience, or they can move into them from outside. Immigration is an important process. It's one of the vital rates that causes changes in population sizes—but one that, for the sake of simplicity, I initially put to one side. Now we see that it's also likely to matter for structuring ecological communities.

I've ignored immigration for too long. It's time to reveal what a vital process it is for the natural world.

Chapter 6

The Silver Y
The Importance of Migrants

Come my friends, 'tis not too late to seek a newer world! . . .
To strive, to seek, to find and not to yield!
— Alfred Tennyson

Silver Y, Camden, London.

J uly 10, 2016. After almost two years of competition between the nations of Europe, France and Portugal faced off in the final of the UEFA European Football Championship, broadcast live from the Stade de France in Paris. It would be a memorable game. Not for the quality of the soccer—it took extra time and 109 minutes for Portugal to upset the hosts and score the game's solitary goal—but for a most unusual pitch invasion. The stadium was descended upon by hordes of moths.

This was not a few stray insects, but a true swarm. Moths were everywhere. They clung to the goalposts and nets. They dotted the corner flags. They fluttered over the players and officials. Portugal's star striker, Cristiano Ronaldo—one of the greats of the modern game and expected to be a major influence on the outcome of the match—had to be stretchered off injured in the twenty-fifth minute. As he waited, distraught, for the medical staff to come on and treat him, one of the moths landed on his eyebrow, as if to drink his tears (some moths do do this; within minutes this one had its own Twitter account). Moths are not partisan, though, and the French players were equally bothered. Media reports from the game discussed the insects almost as much as the football. It gave an insight into what it must have been like in Medford in the 1880s when Gypsy Moths were consuming the town.

Photographs and video from the Stade de France show that the great majority of the pitch invaders that warm July night belonged to a single species: Silver Y. These are remarkable moths.

Fresh adult Silver Ys are a subtle blend of pink and brown, but most of the ones I catch in London are a careworn gray. Even worn specimens retain the distinctive silver letter on their forewing, from which the species gets its English and scientific names. Worn specimens also keep their striking profile—a heavy fur collar sweeping up into a punkish tuft of scales on the thorax, dropping down to two more smaller tufts on its back. When the moth is at rest, the overall effect is of a kyphotic dandy in a moth-eaten fur coat.

It's not its looks that make the Silver Y remarkable, though—it shares the essentials of these with several more strikingly patterned and colorful relatives in the noctuid subfamily Plusiinae, many autographed in similar fashion. Rather, it's the moth's capacity for flight. This is an insect less than an inch from nose to tail, and that tips the scales at not a hundredth of an ounce. Yet it's capable of crossing a continent.

Silver Ys pass the winter as adults around the Mediterranean basin, in Southern Europe and North Africa. In spring, some of these moths head north, following the seasonal flush of resources. The first ones generally reach the UK in early to late May, although the main arrival occurs a few weeks later. Sometimes they appear in their millions—in such numbers that the sound of their wings can be heard as a distinct humming in the fields. They can occupy the whole country, from Kent to Shetland. They arrive hungry, and are a common daytime sight refueling on nectar like tiny hummingbirds. As a result, the Silver Y is one species of moth that is relatively familiar to the general public.

These immigrants come to breed, and they quickly get down to it. Their caterpillars can feed on—yes, you guessed it—a wide range of herbaceous plants, such as clovers, bedstraws, and nettles. They will consume crop plants like peas and beans, too, and can be considered agricultural pests. They enjoy the Northern European summer to the extent that come autumn, the spring immigrants can have quadrupled their population.

Silver Ys don't like the British winter, though. When the nights draw in, it's time for the new generation to head south. As many as seven hundred million of these moths stream back across the English Channel to the continent. You might think that such tiny creatures are simply being tossed on the wind, but they are not. They fly up to altitude—typically more than 100 yards above ground—and if they find the airflow there heading more or less south, they migrate. A tail wind helps, of course, but the moths are active migrants. They adjust their flight path to compensate for drift caused by winds not blowing exactly to the south, steering with an in-built compass. With the wind behind them, they can cruise at 25–30 miles per hour. They might cover more than 350 miles on a good night, and be in the Mediterranean after just three nights of travel. An insect that weighs about the same as a raindrop.

Entomologists can now track flying insects using vertical-looking radars, machines that send a narrow beam of radio waves up into the sky to detect the creatures moving through it. The numbers they record are staggering. A recent study over 27,000 square miles of southern England

and Wales estimated that 3,370,000,000,000 insects—3.37 *trillion*—migrate over the region every year. That's 3,200 tons of insect biomass. The great majority of these are tiny animals like aphids, but "large" insects like the Silver Y still contribute around 1.5 billion individuals to the total, or 225 tons. For context, the thirty million swallows, warblers, nightingales, and other songbirds that head south from the UK each winter tip the scales at about 415 tons. In summer, the insects are basically milling around in the air, but in spring they are generally heading north, and in autumn they are largely heading south. It's not only Silver Ys that migrate.

Bird migration is one of the great natural spectacles, but insect migration is equally spectacular. It just largely goes unseen. There are myriad insects on the move above us at any one time. It's only occasionally that we are confronted with the fact—like on July 10, 2016.

Exactly why moths are attracted to lights is still the subject of debate, but one reason may be that they use the moon or stars to help direct them as they migrate. The lights we put on then override these astronomical cues. A rule of thumb like "keep the moon on your right" can help to steer a more or less straight line, because the moon is very far way. Apply this rule to a street lamp, though, and the result is a flightpath that spirals in to the source. The authorities at the Stade de France had left the floodlights in the stadium on overnight prior to the big game. They inadvertently created the world's largest moth trap.

The Silver Ys added luster to what was generally agreed to be a turgid night of football. How wonderful that such swarms of insects still exist!

Animals like the Silver Y—*long-distance migrants*—are an extreme expression of a feature fundamental to all life: the capacity to move. They are what we imagine when we think of migration—animals that undertake seasonal movements. They breed in one part of the world, and then, because conditions become unfavorable in the breeding grounds for some of the year, they move (migrate) to another area to wait for those conditions to improve. In the northern hemisphere, migrants head south for the winter and then return in the spring. Some fly from the

Arctic to the Antarctic and back every year. Their reappearance is long anticipated and a welcome harbinger of the good times of summer. In other parts of the world, these migrants may be moving south in the summer, or up and down mountains, or between wet and dry areas. In all cases, though, these migrations are more or less predictable seasonal movements, generally of entire populations, to avoid times of hardship. Or perhaps more accurately—because most groups of plants and animals originated in the tropics—to exploit times of abundance.[i]

To an ecologist, though, migrants can be individuals with much more modest ambitions. Migration can describe the daily behavior of animals and plants. Planktonic animals like copepods (tiny pelagic crustaceans) move up and down in the water column, for example. They rise to the surface at night and sink down to lower depths in the daytime, perhaps to avoid diurnal predators or damaging solar rays. Larger animals migrate up and down the seashore with the tide.

We also use the term *migrant* to reference an individual that moves in and out of a population—prefixing with e- or im- to clarify the direction of travel. Most species have different populations, and these populations will by definition be connected by migration. Populations that do not exchange individuals—and therefore genes—with any other populations are reproductively isolated, and they present a case for being considered separate species.[ii] At least they may be on their way to becoming separate. Migration between populations is thus important for both ecology and evolution.

Movement within populations—*dispersal*—matters too, of course. The movement of individuals from their place of birth to a new spot to breed, for example, or from one breeding location to another. It helps in the shuffling of genes. Sometimes, dispersal takes an individual from one population to another.

The ability to move is fundamental to the persistence of all life.

---

i. We'll come to this in the next chapter.

ii. However, this depends on how we define species (see also the next chapter). Here, I would just note that we might define different populations as the same species if they *could* exchange genes, or only if they actually *do*. The evolution of species is a continuous process, and so it can be difficult to say whether different populations are on the way to becoming different species, or have completed the process.

Without movement, individuals would not have new resources to exploit. Populations would not be able to grow. Communities would not diversify. Land would be barren. Even organisms that we consider to be static—barnacles, corals, and lichens, for example—all have the means to travel. It's easy to think of plants as rooted to the spot, but of course they can—and must—move at some point. Indeed, they have a life stage to which movement is integral. A seed that falls in the shadow of its parent tree will be starved of light, water, and nutrients. It will be prey to the same herbivores and diseases that its parent attracts. One imperative of seed design is to produce a propagule that will escape the natal home. Travel is not a luxury, but essential.

Movement is a continuum, from the daily perambulations of individuals to the ebb and flow of entire populations between continents or oceans. It affects processes acting at all levels of ecological complexity, and we ignore it at our peril. Without it, my moth trap would just be an empty box with a light on top—such are movement's impacts on populations and communities. Let's start by thinking about those impacts on populations.

Remember how populations change in size. They grow because of births and shrink because of deaths. Birth and death are guaranteed for all individuals that make it into a population, and a population will grow as long as the former outnumber the latter. Or at least, that's the case for a *closed* population—one for which there is no movement of individuals in or out.

Most populations are not closed, though. They can also change in size because individuals leave them for pastures new, or arrive from those pastures. I ignored these migrants when I introduced models for population growth in chapter 1—the exponential and the logistic—because ignoring them makes life easier. But in the real world, migrants matter. Birth, death, immigration, and emigration are the only four ways that the size of a population can change. Remember from chapter 1:

$$N_{t+1} = N_t + B - D + I - E$$

This equation shows that a population can grow over time even if the number of deaths is greater than the number of births, if there is also immigration. Immigrants can supplement the birth rate. Emigrants have the opposite effect, of course, but if net immigration $(I - E)$ exceeds excess deaths $(D - B)$, the population will increase. Immigration means that populations can grow even in areas that are not really suitable for them—areas where they are dying faster than they are being born.

Those immigrants have to come from somewhere: from another population, where they count under the number of emigrants. To understand the overall effects of migration, we need to think about more than one population. Let's start with two. Let's also assume that those populations do not grow out of control, but instead are each limited to a carrying capacity. The rate at which the population grows is density-dependent—it slows as it reaches this carrying capacity. Logistic growth, in other words.[iii]

If the two populations have the same carrying capacity and the same probability that any given individual migrates, then nothing much changes. Once the populations reach carrying capacity, births balance deaths, and immigrants balance emigrants. The populations swap individuals, but the total number in the two populations stays the same.

That changes if some of the migrants die before they reach the other population. They might head off in the wrong direction, for example, and never complete the swap. Now the number of immigrants to each population is less than the number of emigrants. This has the effect of lowering the population at carrying capacity, because those lost migrants are effectively extra deaths—these balance the birth rate at a lower level. The populations swap individuals, but the total number in the two populations ends up lower than if everyone just stayed at home. This can also happen if there are no deaths *en route*, but carrying capacities differ. More migrants leave the larger population (the probability that any one individual leaves is the same, but there are more individuals); however, they go to an area with a lower carrying capacity. Their loss is not compensated by the smaller numbers heading in the opposite direction. Overall, migration means the species is worse off.

---

iii. See chapter 1 if you need a refresher.

Migration would be a poor strategy if that was the end of it. However, the movement of individuals between populations can also cause species to be better off overall. A large population can feed large numbers of migrants to another, smaller population. If density-dependence is weaker in the smaller population, then immigrants can increase the size of the population above the number that its habitat should support. In effect, the immigrants add to the population faster than the negative effects of density-dependence can remove them—the population overshoots its carrying capacity, but stays overshot. The result can be more individuals surviving in the two populations combined than would be the case without migration!

This effect has another unexpected consequence. The emigrants from the large population decrease the size of that population, because their loss is not compensated by the smaller numbers of migrants heading in the opposite direction. But they also increase the size of the smaller population—that population is raised above its carrying capacity. The actual sizes of the two populations are more similar than you'd expect from the carrying capacities of the two areas. In some situations, the populations can essentially be the same size, even though one area would support more individuals than the other if there was no migration between them. To the casual observer, two areas might look equally good for a species, judging by the numbers living in them, when in fact there is one good area (larger carrying capacity) and one poor area (lower carrying capacity).

What if a developer wanted to build houses on one of the two areas? If you simply tot up the numbers living in each, you might be lulled into thinking that it makes little difference which they choose. Pick the one that is the source of more migrants, though, and the other population would quickly shrink. All of nature is linked, and often in ways that counter intuition.

It could be even worse, of course. As I mentioned earlier, populations can grow even in areas that are not suitable for them—where death rates are higher than birth rates—if net immigration compensates for the excess deaths. We might then have two populations, one a source of emigrants, and one a sink sucking those emigrants in. Without migration, only the source would persist. Picking the wrong location to develop could lead to the loss of *both* populations.

Such *source–sink dynamics* can cause problems when animals are attracted to bad neighborhoods. We might think that individuals would naturally prefer better habitats, but that isn't always the case. Cooper's Hawks live at a much higher density in Tucson, Arizona, than in areas around the city. They nest earlier there, and lay larger clutches of eggs. Unfortunately, the pigeons that form the bulk of the hawk's diet in Tucson pass on a disease—trichomoniasis—to its nestlings, which die at an unsustainable rate. If it wasn't for hawks migrating into the city— perhaps drawn in by a plentiful supply of pigeons—the population there would soon die out.

Moths might be particularly susceptible to "ecological traps" like this thanks to their attraction to light. As we saw in the last chapter, cities generally aren't great for moths. Habitat is thin on the ground, and patchy where it does occur. One thing that *is* plentiful in cities, though, is light. Artificial light at night (ALAN) has substantial negative impacts on the abundance and development of moth caterpillars. Urban light pollution can be amplified by cloud cover so that its effects extend much farther than you may imagine—if you've ever been outside after dark in the countryside, you might have noticed a distant orange glow locating the nearest conurbation. Moths may well be drawn in from good habitats to areas where their populations can only be maintained by the flow of immigrants. Those Gypsys and Uncertains I catch on my roof terrace might not be evidence of species that can coexist with us deep in the hearts of our towns or cities. Rather, they may be individuals trapped in unsuitable areas by a fatal attraction.

Thinking about two populations shows that migration between them matters. It can affect the sizes of the populations, but also whether or not the populations persist. This is becoming ever-more important in the modern world. The reason is that habitats are becoming ever-more patchy.

Patchiness has always been a feature of the natural world. Areas of suitable habitat for a species can be surrounded by others that are anathema. Think about how the environment looks for pond-dwelling organisms—lots of small oases in a sea of earth. Yet if you build a pond in your garden (and if you can, you should), aquatic life will quickly arrive. The Water Veneer I found in my Camden moth trap one July morning is an example. Caterpillars of this moth are aquatic. They feed below the surface of ponds on plants like waterweed and pondweed. They even pupate underwater. Many female Water Veneers are flightless and never leave their natal pond, only coming to the surface to mate. Yet some females can fly, as can males, allowing them to migrate and colonize new water bodies. I can only see one pond from my roof terrace—perhaps my Water Veneer was born there? If not, how far had it traveled?

Small islands in an otherwise inhospitable sea is probably how the landscape looks to many terrestrial organisms, as well. Several species of moth are found in Britain, mainly above the tree line in the Scottish Highlands—the Black Mountain Moth and Broad-bordered White Underwing, for example, that feed on low-growing shrubs like Crowberry and Bilberry that carpet the mountain tops. These summits are their oases in a sea of forest. The Brindled Clothes Moth is mainly found in tree cavities occupied by hole-nesting birds (its scientific name, *Niditinea striolella*, from *nidus*, the Latin for "nest," is a better guide to its habits), where they may feed on feathers. They are apparently reluctant to leave these tiny patches of suitable habitat once they find them. It seems a tenuous existence.

Now humans are contributing to this patchiness through our own impacts on the natural world. Over much of Europe, continuous forests that originally stretched for thousands of miles in every direction have been reduced to woodlands or copses in a matrix of cropland or pasture. Areas under natural grassland or lowland heath have equally been reduced to remnants. Not just in Europe, either—habitat loss and fragmentation are features of the environment worldwide. The smaller patches of habitat that we end up with will support similarly diminished numbers of plant and animal species. Many large natural populations are being repackaged into lots of smaller—and scattered—sub-populations.

And here's the concern: we know that small populations are more susceptible to extinction through the vagaries of chance. Bad luck can cause species to disappear from suitable areas once their numbers drop low. We are increasingly faced with areas of habitat without their typical species, because those denizens have been claimed by demographic and environmental stochasticity.

More could have been taken, though. This is where migrants can fly to the rescue. The presence of a species in one patch of habitat can be a lifeline for the species in another. Migrants matter.

When a population is chopped into many small pieces, the likelihood that any one of those populations goes extinct by chance in some defined period of time is very high. The chance that *all* of those populations go extinct by chance is much lower. Some simple sums illustrate this.

Let's say, hypothetically, that the chance that a population of Water Veneer moths goes extinct in any one year is 0.5—50:50, in other words. Water Veneers inhabit ponds, and in nature these can be small and uncertain habitats. They may temporarily dry up over the course of a summer. If a pond dries up, its population of Water Veneers goes extinct. The pond can refill once the rains return, but the moths are gone.[iv] The likelihood this happens in a year is the same as heads on the toss of a coin. I wouldn't fancy betting my life on those odds—it's a precarious existence. Yet, for the Water Veneer population to survive, the pond has to stay continuously wet year after year—the coin has to keep coming down tails. On those odds, the chance of the pond lasting two years is only 0.25 (0.5 × 0.5, or 0.5^2)—one in four. Ten years and it's down to less than one in a thousand (0.5^{10}). It's basically *guaranteed* that the population of Water Veneers in this pond goes extinct.

The odds that work against the survival of Water Veneers in a single pond work in their favor across multiple ponds, though. If there are two

---

iv. Hypothetically, that is. In fact, many species living in ephemeral water bodies have evolved strategies to survive such dry spells.

ponds, the chance that they both dry up for part of a year is one in four. The flip side is that there's a three out of four chance that one of the ponds stays filled. Ten ponds, and the odds that at least one keeps its water are up to 999 out of 1000. So while it's a precarious existence for the Water Veneers in any *single* pond, they're unlikely to lose—and go extinct from—*all* ponds.

Of course, these probabilities assume that the chance of any given pond drying up is independent of what happens to the other ponds. That may or may not be true—a hot, dry summer may cause all ponds to dry out at the same time. But they also don't factor in the effects of migration between populations—and as we know, migrants matter. A group of populations linked by migration—like the Water Veneers in different ponds—is termed a *metapopulation* (a population of populations). Ecologists have spent a lot of time developing models to understand the dynamics of metapopulations. What these models tell us is that migrants can help populations within metapopulations—and indeed the metapopulations themselves—to persist.

Let's imagine a metapopulation of Water Veneers, where each population consists of the moths living in a single pond. Any of those populations can go extinct, leaving the pond empty of moths. We can assume for now that the chance of this happening does not depend on what the moths do—they can't influence whether or not a pond dries out, say. But an empty pond can be (re)colonized by immigrant Veneers from another pond, and the moths *can* influence this. If most of the ponds have Water Veneers, the chance that immigrants colonize an empty pond is obviously higher than if there are Veneers in only a few of them. How high also depends on how likely Water Veneers are to move from pond to pond—technically, how much each extra pond with moths adds to the likelihood that a pond without moths gets colonized.

Putting all of this together, the proportion of ponds with moths depends on the ratio of the probability that a population goes extinct, and the probability that immigrants colonize an empty pond.

If Water Veneers are more likely to go extinct from ponds than to colonize empty ponds, then the metapopulation will eventually die out. If the reverse is true, then the metapopulation will persist—there will always be moths in some of the ponds. The more that the likelihood of colonization exceeds the likelihood of extinction, the more of the ponds will be occupied by moths. It's like the balance between birth and death rates in a single population—where those rates balance out gives the number of individuals. Here, the balance between colonization and extinction rates gives the number of populations.

If extinction never happens, then all ponds will have Water Veneers. But even if the Veneers are quite likely to go extinct from any one pond, the species will carry on just fine, as long as immigration from other ponds more than compensates for the losses. Populations may die out, but the metapopulation lives on.

In fact, it's also possible that species can influence the chance that populations in a metapopulation go extinct.

Even if the ponds don't dry up, the Water Veneer populations in them are small, and we know small populations are easily lost by accident. Populations dwindling to extinction—through demographic stochasticity, for example—can be rescued from this fate by the arrival of immigrants from other populations. If the extinction rate exceeds the colonization rate, then the metapopulation will still die out in this situation. But if Water Veneers colonize ponds faster than they go extinct, eventually all ponds will have moths. The migrants not only move into empty ponds, but also to occupied ponds, and this *rescue effect* prevents occupied ponds from losing Water Veneers to extinction.[v]

Remember that all models come with a set of assumptions underpinning them—things we accept as true for purposes of the maths. The models that I've illustrated using the Water Veneers make some assumptions that are unlikely to be true in real life.

One of these assumptions is that how the patches are distributed across the environment is irrelevant. Immigrants are just as likely to arrive in an empty patch from a distant patch as from a neighbor.

---

v. Levels of occupation can vary between the two extremes—all or none—if extinction and colonization rates vary over time, which is not unlikely, given the vagaries of the environment.

Another is that immigrants only arrive from other patches in the metapopulation—patches exchange individuals, but only individuals that they produce. A third is that all the habitat patches that make up the metapopulation are the same—the same size, the same quality, no more or less isolated than any other patch. All ponds are created equal.

The reality is that some habitat patches are going to be better than others for any given species. These will usually be the big ones. Once again, size matters. They will be home to larger, healthier populations, and likely to be sources of emigrant individuals. These emigrants may end up as immigrants to sinks, or just to smaller, more precarious populations in more-marginal areas—rescuing these populations from extinction, or recolonizing patches where extinction has occurred. There might be a freshwater lake supplying Water Veneers to gardens in the neighboring suburbs—the chain of large ponds on Hampstead Heath may do the job where I live. Ecologists being terrestrial creatures, we more usually picture a large area of mainland habitat pumping emigrants away to offshore islands in a literal sea. Islands and their relationship to the mainland have been fundamental in the development of ecology—and evolution, too, of course, not least through Darwin's visit to the Galápagos. I'm writing from an island myself, and looking forward to the imminent arrival of Silver Ys (and other migrants) from the continental mainland to which we are vitally connected.

For metapopulations, the existence of a large "mainland" population results in a constant rain of immigrants to the other, smaller habitat patches—be they "real" islands or islands of habitat. This removes the task of producing enough immigrants to keep the metapopulation going from the constituent populations. It has the effect of guaranteeing that the metapopulation survives—even if it's highly likely that populations go extinct, and highly unlikely that immigrants arrive. Because the mainland is always producing emigrants, some of the island patches will always be occupied. How *many* patches are occupied still depends on the balance between the rates of extinction and colonization—but as long as there are immigrants there will be occupation. Better not lose the mainland population, though . . .

It may be a little clearer now why ecologists started out trying to understand how populations change in size by ignoring migration. The situation can be complicated enough just thinking about births and deaths. Adding immigrants and emigrants into the population mix inevitably adds layers of complexity to their dynamics. As usual, ecologists have tried to make their lives easier by breaking the problem down into more manageable parts. Metapopulations, sources and sinks, islands and mainlands—these are all ways of thinking about how populations of plants and animals exist, and persist, in the landscape (or waterscape) once we grasp the nettles of immigration and emigration. In the real world, of course, species do not recognize these neat distinctions. But the models are still useful in guiding our understanding.

Take the Oak Carl, a tiny, golden-brown moth that mines the leaves of—you guessed it—oaks.[vi] To the Carl, oak trees are like islands in a hostile sea, and like islands they are scattered effectively at random across that sea. Oaks also vary in size. Larger trees have more leaves, and more leaves means more opportunities for the moth. It's quite straightforward for scientists to spot and count the blotchy patches of discoloration that Carl caterpillars cause as they tunnel between leaf surfaces.[vii] This allows them to follow the ebb and flow of Carl populations across different oak tree islands. That's exactly what Sofia Gripenberg and her colleagues did on a small island off the southwestern coast of Finland.

Large oak trees are more like moth mainlands than islands. They have lots of leaves and lots of moths. The large moth populations on these trees more or less never go extinct. They are sources of Carls, pumping out emigrants into the wider landscape.

It's different on small oak trees (for scale, one up to about twelve feet tall). With relatively few leaves at their disposal, the Carl populations on these trees are small, too, and precarious. They often go extinct.

---

vi. And, as you probably didn't guess, Sweet Chestnut. As with the Brindled Clothes Moth, the accuracy of English names is not guaranteed.

vii. Straightforward, that is, but quite laborious—because field ecology is.

If it weren't for immigrants from the large "mainland" oaks, Carls would not be found on these small "islands." These populations are sinks. Which islands have Carls varies from year to year thanks to the element of chance inherent in whether or not immigrants find that oak. Populations maintained by colonization in the face of repeated extinction is a classic feature of metapopulation dynamics.

It's not just size that matters—*where* the oak trees sit with respect to each other also matters. If large trees are isolated, they still pump out emigrants, but those moths mainly disappear into the void. These trees are sources, but not sources that are of much benefit to the sink populations that need their moths to stay afloat. Remember that assumption of metapopulation dynamics that how the patches are distributed across the environment is irrelevant? It isn't.

Metapopulation dynamics break down if trees are too isolated, but do so too if they are too clumped. When oaks are packed together, they stop being separate patches for the Carls, linked by migrants, but instead become essentially one big population. Think of these clumps as large, irregular islands, rather than archipelagos. The Carls on any one tree in the clump are as likely—more even—to have grown up on a leaf from a different tree as from a leaf on the same tree.

A single species of moth feeding on a single species of plant, on a single island. A simple system, but one with parts that fit into many of the categories our models attempt to define. Some of the Carl's populations are islands, some are mainlands; some are sources and others are sinks. Some of them form metapopulations, some are just large, patchy populations.

And this is just a snapshot in time. As the small oaks grow and the large ones eventually die, as acorns sprout, these islands and mainlands of habitat for the Carl will themselves migrate over the landscape. Sinks will become sources, and sources will disappear. The Carl will follow along behind, buffeted by those four basic processes: birth and death, immigration and emigration.

Not all species follow changes in the patches they depend upon, though—some drive them. The Cinnabar moth we met in an earlier chapter is one example. It turns the relationship to its food plant— Ragwort—on its head.

Recall that Cinnabar caterpillars can chew Ragwort plants to their stems—and indeed beyond. The intensity of herbivory can be so high that it actually helps to drive the extinction of populations in the Ragwort metapopulation. The Cinnabar literally eats away the island on which its life depends, putting its own survival at risk, too. There is enough dispersal of moths between Ragwort populations that the Cinnabars themselves are probably a large, patchy population. But without the Ragwort, that population would not last long.

Fortunately, metapopulation dynamics come to the rescue.

Ragwort populations can reestablish in habitat patches, thanks to a combination of seeds lying dormant in the soil, and seeds blown in from source populations that never go extinct. Their recovery is aided by a specialist parasitoid of Cinnabar Moths—the wasp *Cotesia popularis*. The wasp parasitizes more Cinnabar caterpillars when caterpillar numbers are low, keeping caterpillar numbers low for an extra year. The delay that the parasitoid imposes on Cinnabar Moth population growth gives Ragwort patches extra time to recover. The moths play whack-a-mole with their food plant as it colonizes habitat patches, but metapopulation dynamics keep the Ragwort going, despite the impact of the moth. That's lucky for the Cinnabar, which couldn't survive without them.

Migration matters for populations. Yet we know that populations do not exist in isolation. Different species live side by side in the same area, in communities. For that reason alone, migration will matter to communities, as well.

We saw this in the previous chapter. Communities have context— they are connected to the wider environment. Individuals can be born into communities, or they can move into them from elsewhere. This is the basis of Neutral Theory, but this underlying fact still applies even though fundamental assumptions of that model—notably that all species are ecologically equivalent—are not true. It is most clearly demonstrated in a feature of communities that we did not really come to grips with in the last chapter.

One of the problems in trying to understand how ecological communities are structured is that they are protean. We want to pin down what "the" community is, but it constantly changes. Every year, different species come to the moth trap. New species get added to the trap list. Some of those new species stay—have they just moved into the community, or were they always there but just missed by the trap? Other species stop appearing in the catch, apparently having succumbed to local extinction. Changes over time make it hard to define the community we want to understand. But much as we'd like to, we can't ignore them. Those changes in community membership— *turnover*—are also an integral part of what we need to understand. Like the Vine's Rustic and the various species of Footmen. They show us that community membership is dynamic. Species colonize communities, and go extinct.

You can think of the turnover of species in a community in a similar way to the turnover of individuals in a population. At carrying capacity, when a population is in equilibrium, births and deaths balance. The *number* of individuals in the population is constant, but their *identity* changes over time. And the same is true for species in a community, when colonization and extinction balance. Of course, in a community, it's possible for species that go extinct to reappear at a later date. But when individuals in a population die, not so much.

Colonization and extinction—sound familiar? The balance between these two processes influences the distribution of populations across a patchy environment. It can do the same for the distribution of species— and so the dynamics of communities as well. We owe this insight largely to the work of two of the most influential ecologists of all time, Robert MacArthur and E. O. Wilson.[viii] Their classic model of the process is known as the *Equilibrium Theory of Island Biogeography*—Equilibrium

---

viii. Wilson coined the term *biophilia* that we first met in the introduction to this book. MacArthur made seminal contributions to the development of ideas on density-dependence versus density-independence, the nature of competition, life history evolution, causes of species diversity patterns, and the structure of species geographic distributions. More or less every topic covered in this book, in fact. His impact on ecology is all the more remarkable given his tragically early death at the age of just forty-two.

Theory, for short. It was inspired by the biodiversity of islands. It's a good place to start in understanding how immigration and extinction combine to affect the composition of ecological communities.

As we saw for metapopulation dynamics, islands can come in many forms. It's probably easiest here to start by thinking about literal islands, though.

Let's begin by assuming that islands are blank slates, completely free of life. Many islands actually start out like this. We can see their genesis in Iceland, where they bubble out of the sea as molten lava. Likewise Hawaii. Sometimes islands revert to blank slates after their life is wiped away. This happened to those that made up the Krakatau group in Indonesia in 1883, when the volcano on the largest of them erupted with catastrophic force. Around 36,000 people were killed by the blast and subsequent tsunami. It also wiped out all life on Krakatau—not a single organism survived there. Yet, within a year, the first land animal—a spider—was spotted among the devastation. A few months after that, the first blades of grass appeared. Life was quick to find a way back in, thanks to migrants. Krakatau sits in the Sunda Strait between the Indonesian islands of Java and Sumatra. Both of these large landmasses were likely to be rich sources of the immigrants that started the process of recolonization.

The arrival of immigrants brings individuals to islands, but it also brings species. Each one that arrives adds to the richness of the community that lives there, and the community grows. Each species that arrives also reduces the number of species that can colonize in future—because they're already there. The rate at which new species arrive therefore drops. In theory, species *could* keep colonizing until the island was home to every species in the mainland source—at which point the rate at which new species arrive would, of course, drop to zero. This doesn't happen, though, and the reason is the counterbalance provided by extinction. Extinction removes species from the island. How many species end up sharing the island hinges on where the opposing forces of colonization and extinction level out.

The more species that colonize an island, the more pressure they are going to put on whatever resources that island has to offer. Resources are everywhere finite. Keep adding more species, and that finity will be divided into ever smaller amounts. As we know, small populations are more likely to go extinct just because of the vagaries of chance. As the number of species goes up, so too does the rate of extinction, as the population size of each (on average at least) goes down. Eventually, the rate at which species are being lost equals the rate at which they are being added. Extinction equals colonization, and the richness of the island community stops growing. It has reached the equilibrium from which MacArthur and Wilson's Theory takes its name.

Small islands will have fewer resources, and so fewer species will be able to grab a share before extinction and colonization rates balance. Larger islands will have more resources, and will end up with more species. Equilibrium Theory predicts that species richness will increase with island size.

It does. More species live on larger islands than on small. This is probably the most robust pattern in all of ecology—called the *species-area relationship*. For example, Great Britain has around 2,400 species of moth recorded, Ireland just 1,400. The Isles of Scilly off the western tip of Cornwall, perhaps 500. The same pattern is found in more or less every group of organisms counted on more or less every group of islands. A very rough rule of thumb is that a tenfold increase in the area of an island doubles the number of species. When it comes to islands—or indeed any piece of habitat embedded in inhospitable surroundings—size matters.

Size is expected to matter, but it's not the only thing. Isolation should, too.

Hitting a dartboard isn't that difficult if you're throwing darts from the oche—ninety-three inches from the wall is standard. It's much harder from thirty feet, and near impossible from a hundred. Immigrants are the darts, and the island their board—their chances of hitting it decrease the farther away it is. The lower immigration rate to more distant islands means that extinction will balance out richness at a lower level. Once they reach equilibrium, distant islands should have impoverished communities compared to islands nearer the mainland.

They do. This pattern is less universal than the species–area relationship, but it is still a common finding. Contrast the moth fauna of Great Britain, sitting within sight of the European mainland, to that of Iceland: fewer than a hundred species, mainly of European origin. Admittedly, Iceland is only about half the size of Britain, but that number of species is far lower than would be expected from area alone. That a significant proportion of the species that do occur on Iceland are long-distance migrants—like the Silver Y—attests to the difficulty of hitting a distant target.[ix]

MacArthur and Wilson's original idea was that size matters for the rate that species go extinct, and isolation matters for the rate that species colonize. The reverse can also be true.

Larger islands make bigger targets for immigrants to hit. It's easier to hit a dartboard thirty feet away if it's thirty feet across. Those Silver Ys heading north across the English Channel might miss the square mile of Tresco in the Isles of Scilly. They're unlikely to miss the 94,000 square miles of Britain. More immigrant species hitting a larger island will balance extinction at a higher level—giving a richer community. The ease of hitting a closer island can also save species from extinction, if the immigrants bring reinforcements to a population teetering on the brink. Immigrants plus births can exceed deaths, when births alone would not. This is the rescue effect we met earlier. Size and isolation can both affect colonization, and can both affect extinction.

This brings us back to the problem of the changing nature of communities. Change creates difficulties for the question of how communities are structured, but is also part of the answer. If the number of species living on an island (or other area) is a balance between the arrival of species, and their extinction, then we would *expect* communities to be protean. The identity of species *should* change over time, even if the number of species in the community is at equilibrium. We expect turnover.

How rapidly the species in an island community turn over depends on where colonization and extinction balance. This depends on their respective rates.

It's again analogous to births and deaths in a population. When the death rate is low, the population only stabilizes if the birth rate is

---

ix. And several of the other Icelandic species probably hitched a ride with humans—the Common Clothes Moth, for example.

low, too. There's little turnover of individuals when few die. There's a lot of turnover when individuals are dying at a high rate. Think of the human and rat populations of London—the identities of the rats (higher death rate, higher birth rate) change more quickly from year to year. Populations can be the same size but have very different rates of turnover.

The same is true for communities on islands. Low rates of immigration and extinction can balance out at the same number of species as high rates. In this way, a large, distant island (low immigration, low extinction) could be home to the same *number* of moth species as a small island closer to the mainland (high immigration, high extinction). But we would expect the *identities* of the species on the latter to change much more rapidly.

Explaining turnover was an important motivation for MacArthur and Wilson's models. Communities do not assemble into immutable collections of species: membership changes. The expectation of turnover is a strength of their model. But it's also a weakness.

Colonization and extinction do affect the species richness of islands, but to understand the development of these communities properly, we also need to factor in the *identities* of the migrants. Equilibrium Theory treats all species as if they were the same.[x] But of course, they are not. The Goat and the Codling were not created equal. Neither were the Silver Y and the Oak Carl. Communities *do* turn over, but *not* as we expect if species are treated as equivalent and interchangeable. Identity matters.

Krakatau is a nice illustration. Following the eruption, the first plants to colonize were ferns and grasses. The spores and seeds of these species are small—they can be easily caught up on the wind and blown to new opportunities. They are good colonizers, but this comes at a cost—they trade off abilities as colonizers for abilities as competitors. Recall that no species can be good at everything, so trade-offs are inevitable.

Eventually propagules of shrubs and trees arrived on Krakatau. These plants trade off in the opposite direction. Their large seeds find

---

x. Like Neutral Theory, which Equilibrium Theory inspired.

it harder to travel long distances,[xi] but once they do arrive, the large stores they carry allow them to produce long, strong shoots and roots. They are strong competitors. Grassland was put into retreat as shrubs, and then forest, spread over Krakatau. This sort of succession will be familiar to anyone who has watched nature take over an abandoned garden. Species colonize, and—it's true—species also go extinct. But colonization and extinction are not just random. Extinction rates rise among the grassland species as shrubs take over. Then shrubs decline as forest moves in. New species still arrive—albeit at a slower rate, as Equilibrium Theory predicts—but most of them fail to establish. The extinction rate among the residents does not go up—it drops, too. Once the canopy of the forest closes, it's hard to dislodge the incumbents. It's the same priority effect that we saw in the previous chapter. Possession is nine-tenths of the law.

Animals cannot make their own food, and so depend on the produce of others. Their communities inevitably depend on the opportunities provided by—ultimately—the plants and other autotrophs that colonize islands. Those day-flying moths we know as butterflies—and likely the nocturnal moths, too—followed the vegetation changes on Krakatau. A burst of immigration, followed by turnover as forest species replaced those of grassland. Then not much in the way of immigration or extinction, nor much in the way of turnover.

Identity also determines *which* animal species get the chance to exploit these insular opportunities. What habitat a species likes is only part of the story. When the opportunity to exploit an opportunity depends on being able to get to it, the ability to travel is key. And species differ greatly in mobility.

Size matters here, too. For moths, species with larger wingspans are able to disperse farther, and are more likely to reach islands. Around six out of every ten species in the UK is a micromoth. In Iceland, six out

---

xi. They often rely on animal assistants. For example, the spread of oak trees across Northern Europe as glaciers receded at the end of the last Ice Age was speeded by European Jays carrying acorns off to bury and eat later. Forgotten acorns (or those reprieved by the death of the Jay) then had the chance to break new ground. Millions of nutritious nuts and fruits get eaten, but enough survive to germinate for vast forests to grow.

of every ten species is a macro—and many of the micros clearly arrived as stowaways. Robust, well-muscled, fast-flying Noctuid species—like the Silver Y—are more likely to be found in insular habitats than their more willowy Geometrid cousins. So too are species that are abundant in the source community, because migrants will be more likely to come from them by chance alone.

Getting to the opportunity presented by a new location is one thing. Being able to exploit the opportunity is the next. Not being picky about diet helps here. Arriving to find that there's no food for you—or more importantly, your offspring—will stop colonization in its tracks. Those moth species that can develop on (yes!) a wide range of herbaceous plants are likely to find something to their taste. Especially if the island or habitat is in the early stages of succession, but also in the understory after trees take over. It's why I find so many species like this in the isolated patches of city habitat sampled by my moth trap in London.

Species with low mobility and specialized diets find it difficult to persist in isolated habitats. Patches are less likely to have their food, *and* they find it harder to move to another one that might. This affects patterns of extinction, because species that *can* persist in isolated locations, *do*. Resources determine which species can move in, and it's then hard to dislodge those that establish. Remember the previous chapter: later immigrants inevitably arrive in low numbers compared to the host of residents they have to square up against, and will find it difficult to grow their populations away from the marginal levels where demographic stochasticity is a danger. Species can move in if resources change again—like the Footmen in Britain after the resurgence of lichen—but they need that new opportunity. Without it, the tenants continue to sit.

Ecology is underpinned by a few simple truths. One of these is that movement is fundamental to the persistence of all life. I'd get nothing at all in my moth trap if it weren't for movement. And not just for the prosaic reason that animals have to fly into it.

Immigration and emigration sit alongside birth and death as the basic processes underpinning the generation and maintenance of all ecological systems. Their interplay within and between populations determines whether populations live or die. Our models help us to understand these dynamics—and why movement is so important—although no single model captures them. Nature is too complicated. All models are wrong, right? True. But some are useful. Models of metapopulation dynamics have been hugely influential, along with Equilibrium Theory. None of them are perfect depictions of the workings of nature, but they do capture key elements. Some of those elements turn out to be very important indeed.

The first is just how much immigrants really matter. (Emigrants, too, of course—every immigrant is also an emigrant.) Much of the world would be home to no life at all without them. The Hawaiian Islands, Iceland, Krakatau. The Galápagos that so inspired Darwin. The life he found there influenced his theories on evolution, but these volcanic outcrops would be bare rock had not propagules arrived from elsewhere. Most of the UK was under ice just 12,000 years ago. This means that most of the species with which I share my island home colonized as the glaciers retreated. Just as my ancestors did.

And immigrants matter not only for true islands, but also for islands of habitat in a hostile matrix, like gardens in a sea of concrete and brick. They keep marginal populations going, maintaining diversity that would otherwise be lost. They bring Water Veneers and Silver Ys to a roof terrace in urban London, and with them the joy of the unexpected in a moth trap. They show that understanding the contents of the trap is not just a question of looking over the fence to see what's in the next garden or field. Some of the moths that touch down in my trap took off from France. From Cristiano Ronaldo's face to my roof terrace, perhaps. Your local ecology is connected to the wider world in unexpected ways.

Metapopulation models and Equilibrium Theory also identify extinction as a fundamental process driving patterns of diversity. It might seem odd, but extinction is to a population as death is to an individual— inevitable, and something that cannot be ignored. The eruption of 1883 was a dramatic demonstration, the sudden termination of all animal, plant, and other populations on Krakatau. But most populations end with a whimper rather than a bang.

Krakatau also highlights the fact that, while extinction marks a loss, it also represents an opportunity. Other species can take advantage of the vacancy. New immigrants have a chance to establish, and turnover occurs. Communities change. Change is the only constant. But increases in extinction rates will have consequences—and thanks to the importance of movement, not just for the populations directly lost. Concreting over Hampstead Heath would be a catastrophe for the diversity of moths attracted to my roof terrace in Camden, even though none of that destruction would be visible from it. What of destruction farther afield? Every little piece of habitat lost has incremental effects that may be felt surprisingly far away.

Our models show that identity matters, too. The Codling and Goat are not the same, and one cannot blindly substitute one moth for the other. The Stade de France was stormed by Silver Ys, not Footmen.

It's not enough to treat species as interchangeable bricks in the construction of communities, because traits determine which species can make it to patches of habitat, and which species persist once they do. Traits influence metapopulation dynamics in the same way. Species are not equal in terms of their abilities to colonize patches of habitat. Species are not equal in terms of their likelihood of extinction. That has consequences for how communities change.

There is one more element of nature that immigration tells us will matter, though, and it's a big one. It's a feature of Equilibrium Theory, albeit one I've largely ignored to this point. It's the *source pool*.

Remember that each species that colonizes a patch of habitat—island or other—comes from somewhere else: the source of immigrants. For Krakatau, it's Java or Sumatra (perhaps via Sebesi). For the UK, it's continental Europe. Each colonist reduces the number of species that could potentially colonize in the future—but the number it reduces it from is the richness of the source pool. The more species in that source pool, the more potential colonist species.

This really matters. The species on an island, or in a community, are a subset of those in the wider environment. The richer that wider environment, the richer the community represented by that subset will be. I'll never catch more species if I run my moth trap in Reykjavik than I have done in London, because Iceland just doesn't have that many moth species. Conversely, I can only wonder at what it must be like trapping on Poring Hot Spring in Borneo, where twenty-four nights of trapping across nine months in 1999 yielded 1,169 species of macromoth.[1] More than the entire macromoth fauna of the UK—at just one location (it's little consolation that they were running six traps). The richness of localities depends fundamentally on the richness of the region in which they are embedded.

So what then determines the richness of the region? The short answer is evolution. For the long answer, turn to the next chapter.

Chapter 7

The Poplar Hawk-moth
Diversification and What Drives It

From so simple a beginning endless forms most beautiful and most wonderful have been, and are being, evolved.
— Charles Darwin

Poplar Hawk-moth, Devon.

169

Three years to the week since my desire to own a moth trap crystallized, I was again with a group of undergraduates at an FSC center. It was the fortnight of my course on field ecology, introducing the students to methods of quantifying biodiversity. This year the trip was different, though. We were at Blencathra in the Lake District, not Kindrogan in Scotland, now sadly closed to field courses. It was 2021, so a suite of Covid-19 protocols were in place. And the sampling was different, too. Running a light trap was not just a pleasant diversion from the main business of the course. This year, I'd made sure that catching moths was very much to the fore.

To understand how the natural world works, we need to know what's there. For several years, the field course students have carried out pitfall trapping. It's a simple but effective method of sampling one community—ground-dwelling arthropods. Dig traps in the evening, and come the morning some of the animals using that habitat will have dropped in. It's not without shortcomings, though. From the perspective of teaching, the main one is that the kinds of species the trap catches— beetles, bugs, flies, collembola, spiders, ticks, mites—are hard to put a name to with any great certainty. The range of taxonomic groups caught does not help: it's difficult to be proficient in the identification of multiple highly diverse lineages. Names matter. If we can't put names to the animals we catch, we can't even calculate basic metrics like the number of species we've caught.

Naming is much less of a problem for a moth trap. Most of the individuals caught are identifiable to species with a high degree of certainty—or the macromoths are, at least. So in 2021, the students set pitfall traps as usual, but as a backup. Plan A involved a clutch of compact actinic Heath traps. Having spent three years engrossed with Britain's moths, learning about their biology and how to identify them, it was time for me to pass on some of that new knowledge to the next generation. Or at least that was the plan. Would the moths play ball?

I was optimistic. The weather forecast looked decent. The night should be mild, the chance of rain—so often a dampener on days (and nights) in the Lake District—low. We retired to bed the first night of the trip leaving Blencathra dotted with light.

I'd agreed to meet the students at 7:00 a.m. to show them how to approach, open, and empty a moth trap. But like a child on Christmas morning, I couldn't wait that long. I was out an hour ahead. Just to check one of the traps. Just to find out whether the presents I wanted had been delivered. As I approached, even from several yards away I could see that the day was going to be a good one—we would have no need for Plan B. Sitting on the outside of that first trap was a Poplar Hawk-moth.

Hawk-moths are insect megafauna. This description is apt in part for their size—not all are giants, but the Privet Hawk-moth is the largest British resident species and has the wingspan of a small bird—but mainly for their charisma. They really are mega!

Hawk-moths are consummate fliers. Many feed as adults by hovering in front of flowers, zipping rapidly from bloom to bloom and drinking by unfurling their long proboscis. The "hawk" in their name is thought to derive from comparison with the Kestrel,[i] that archetypal hovering bird. The Hummingbird Hawk-moth nectars by day, and is frequently mistaken for one of its avian namesakes (although the moth is an Old World species, while hummingbirds are restricted to the New). Other diurnal hawk-moth species mimic bumblebees, developing their disguise to the extent that their wings are transparent, like those of their models. The rare migrant Convolvulus Hawk-moth only has to raise its wings a little to reveal a pair of crimson spots on its thorax, giving would-be predators a demonic glare. Elephant and Small Elephant Hawk-moths (named for their trunk-like caterpillars) are startlingly bright pink and gold, of an intensity in parts so vivid as to seem unnatural.

The Poplar has none of the obvious attractions of these other hawk-moths, but it exudes a charisma all its own. It can capture hearts, as it did for moth fanatic and author of the passionate paean *Much Ado about Mothing*, James Lowen. He describes it as his ur-moth, "huge, glorious, and utterly wrong."[1] Its typical pose is unusual as moths go, hindwings partly protruding forward of the forewings to give a boxy and broken outline. While largely a sober mix of gray and brown, it can flash those

---

i. Technically, Kestrels are falcons, not hawks, but *hawk* was a more generic term when the moths were first named. (In fact, falcons and hawks are not even that closely related.)

hindwings if disturbed to reveal startlingly deep-red patches. It startles only slowly, though. Like most hawk-moths, it's rather chill for an animal with such a heavy fur coat. It seems to have few qualms about being passed from hand to hand around a group of curious students.

The Poplar Hawk-moth is also relatively common, which feels wrong. Something so magnificent ought to be rare, but we caught them every day we ran the traps at Blencathra. In this case, familiarity bred love, not contempt. If anything was going to convince novitiates of the joy of moths, catching Poplar Hawk-moths was likely to be it. And indeed, it wasn't long before students were being photographed in the classic moth-er's pose, hawk-moth on nose. Portraits snapped, they threw themselves into the task of naming the other moths with enthusiasm.

Hawk-moths are mega, but also a nice, easy introduction to the challenges of moth identification—large, well-marked, and distinctive. Unfortunately, they trade off charisma for diversity. Only eighteen species have been recorded in the UK. Half are permanent residents, two are migrants and temporary residents, and the remainder visit as scarce or rare overshoots from the continental mainland. That leaves 850 or so other British macromoth species that generally require a bit more work to name. Quite a lot of work in some cases.

The greatest proportion of these are classified in the family Noctuidae. We've met them already in the Silver Y and Uncertain, but Britain is home to more than 350 species of their ilk. Another 300 or so species are Geometridae, colloquially known as loopers or inchworms from the distinctive gait of their caterpillars. In the UK, many of their common names end with Carpet, for their woven patterns, or Pug, after the drooping jowls of the small dog. Weight of numbers alone raises the identification challenges within these two families, and the adults of many require attention to small and subtle differences in markings for separation. A few differ consistently only in the structure of their genitals. Identification is a bit easier for adults of Erebidae (around ninety British species, including Tigers and Footmen), Notodontidae (around thirty species, including the Buff-tip and various Prominents), and other macromoth families, most of which are represented in Britain by around twenty species or fewer.

The students put names to seventy-eight macromoth species in the nights we trapped at Blencathra. A little shy of one in ten of all British species. There were some real beauties among them—the delicately pink-spotted Peach Blossom, and iridescent gold Burnished Brass, for example, both much anticipated first records for me. But nothing topped the excitement elicited by that first Poplar Hawk.

What we call a thing depends on the language we speak, or where we learnt to speak it.[ii] The United States and United Kingdom may be two nations divided by a common language, but there are fundamentals we share, and share with all humanity. Names anchor our experiences and waymark our narratives. They also let us catalogue, and catalogues allow quantification. Having put names to the moths in a trap, who would not then immediately tot up how many different names they have? No scientist, for certain! And that brings us back to one of the most fundamental questions in understanding the contents of a moth trap. What determines how many species we catch?

I've addressed this already when thinking about the ecology of communities. The locality around the trap certainly matters. The grounds of Lewtrenchard Manor harbor more species of moth than the gardens overlooked by my London flat. Habitat quality and quantity both matter for this. Concrete and brick do not grow moths. A population living on a small patch of suitable food is open to the vagaries of chance. Small populations tend not to last long when luck is against them, be it bad weather or accidents of birth. Competitors and predators compound these problems.

But a community's immediate surroundings only tell part of the story. The question of species number is one that requires us to think more broadly. Globally, in fact.

No locality exists in isolation. The gardens in my street, the grounds of Lewtrenchard—they are embedded in the wider environment. My

---

ii. This is why we give every species a single scientific name, so that we know what we're talking about regardless of our mother tongue.

local moth community is part of a metacommunity—a community of communities—whose membership is crucially important to the contents of the trap. As we saw in the previous chapter, migration is a fundamental process that, with birth and death, structures the natural environment. Small populations that would otherwise be lost to chance can be maintained by migrants from elsewhere in the metacommunity. Small islands import richness from the species pool. The metacommunity comprises more individuals than the local community, and so almost certainly more species. Immigration from it can elevate and maintain local species richness. Richer metacommunities give us richer local communities. And more species in a moth trap.

This of course leads us inevitably to the next question: What determines the richness of the metacommunity? To address that, we need to understand how species richness varies at broad scales across the environment, and why. We need to think about the processes that generate species.

Hawk-moths are mega, but in this they are multiply exceptional. Most moths are small and relatively insignificant—barely noticed by people who don't run a trap.

We have names for around 140,000 moth species (160,000 if we include butterflies, which of course we should), but for most of these, names are pretty much all we *do* have. And moths are a relatively well-studied group. Recall how little we think we know about parasitoid wasps, for which the vast majority of species probably don't even have names. This is all highly problematic for understanding broad-scale variation in species numbers. For that, we need to know not just what species there are, but also *where* species live. And for most insects—indeed most species, full stop—we don't. They're simply too small and too numerous to map their distributions in any sort of detail.

On top of that, what knowledge we *do* have is likely to be highly biased to those parts of the world where most taxonomists and scientists live—mainly temperate northern latitudes. Trying to understand broad patterns in species numbers is complicated if we don't know most of the species present in some regions, like the tropics. It's the same problem we encountered in trying to understand local community structure, but on a grand scale.

For these reasons, most studies on patterns of variation in animal species numbers across the globe come from a few well-known groups. As we widen our perspective, we rely on ever smaller fragments of the whole picture. In this case, those fragments are the charismatic megafauna, primarily birds and mammals.

Birds and mammals have big advantages over all other groups when it comes to understanding why some areas have more species than others. They get more attention from amateur and professional naturalists and ecologists, and so we think we know most of the species that exist.[iii] We know a lot about the ecology of individual species. We also know pretty much where they live: we have maps of the geographic distributions of more or less all 15,000-plus known species, to a relatively high degree of accuracy. This means we can identify areas that are rich and poor in species.

On top of that, we have a pretty good idea of their *phylogeny*—the "family tree" that describes the evolutionary relationships between all of the different orders, families, genera, and species of birds and mammals that we've so far named. This is the result of a stupendous amount of work spelling out and comparing the genetic codes of species, to find out who is related to whom—paternity tests on a grand scale. It tells us that falcons are actually not closely related to hawks, but to the long-legged South American seriemas. And that the moth-eating nighthawks are more closely related to hummingbirds than to owls. Knowing phylogeny means that we can look for differences in the rate at which species have diversified over the vast expanses of evolutionary time. This helps us identify the causes of this diversification.

While we know most about birds and mammals, we are not entirely clueless about smaller species, even some insects. Hawk-moths are megafauna, too—almost honorary birds. For the 1,000 or so species living in the Old World, at least, we know pretty well where they're found. This allows us to map regions that are rich in hawk-moths, and

---

iii. Though we are still discovering new species every year.

regions that are poor, over a significant proportion of the land area of the globe.

What these maps show us is that when it comes to species numbers, different parts of the world are far from equal. There are very much haves and have nots.

The best place to find large numbers of hawk-moth species is the wet tropics. Southeast Asia is particularly rich—some 100 × 100 kilometer areas in this part of the world can be home to more than 175 species. That tops the number found in the whole of the continental United States and Canada combined. It's roughly ten times the number that have been recorded in the whole of the UK, but in an area a little larger than Cyprus.

A finger of notably high hawk-moth richness stretches out northwest from Southeast Asia, along the southern slopes of the Himalayas. Richness is also high through the Indonesian archipelago and into New Guinea, but it seems that hawk-moths have struggled to cross the Torres Strait into Australia. Or at least those that did make the crossing have not then diversified into many species. The wet tropical areas of Queensland aside, Australia is rather depauperate in hawk-moths.

Another hawk-moth hotspot is the central belt of sub-Saharan Africa, through the rainforests of West and Central Africa. Richness peaks here in montane East Africa—the mountains of the Albertine Rift in Uganda, Rwanda, Burundi, and eastern DRC. The wooded savannahs and coastal forests of East and Southeast Africa are also good locations for hawk-moths, as is the northern half of Madagascar.

Richness tends to drop off as we move away from the wet tropics. The drier areas of Southwest Africa and Central Asia, and the cooler areas of Europe and the Russian taiga, are home to many fewer species. At least there is *some* diversity here—enough to enthrall students on a field course to the English Lake District. The arid regions of western Australia, and the tundra of northern Russia, have barely any.

That's hawk-moths, but I could have written more or less the same for the species richness of birds and mammals.[iv] Each has wrinkles on the hawk-moth pattern, but in general, animals like it hot and wet. The

---

iv. And also for that group of day-flying moths that we call butterflies.

presence of mountains jacks up richness further. Southeast Asia and the southern Himalayas, the Congo Basin, and the Albertine Rift—all are hotspots for bird and mammal species, too. The same patterns hold true in the Americas, as well. There, the Amazon and Atlantic rain forests harbor high species richness, but numbers peak most dramatically in the Andes. A 100 × 100 kilometer square area in the mountains of Colombia, Ecuador, or Peru can be home to more species of breeding birds than the whole of Europe.

We don't know for sure whether the megafauna are representative of the rest, but it's a reasonable assumption, at least in broad terms. If you want more species in your moth trap, you need to be closer to the equator, and prepared to get rained on.

Of course, that simply begs the question of *why* we see more species in the wet tropics. Ecologists have been chasing the answer to *that* question since the first observation of the pattern by polymath Alexander von Humboldt at the start of the nineteenth century. The answer is, as usual, complicated.

In general, numbers of species decline as one moves from the tropics towards the poles, but that simple gradient masks a lot of variation. Talking about "the tropics," wet or otherwise, or "the temperate zone," masks the fact that these latitudes are really a mosaic of biogeographical regions, each with their own evolutionary histories and distinctive floras and faunas. Some tropical regions are richer in species than others. We see variation as we move east to west across continents, as well as south to north. Richness likewise varies as we scale mountain slopes, particularly in major ranges like the Himalayas and Andes. They have fewer species at the top than the bottom, though often the most species are found somewhere in between. We need to be able to explain this variation, too.

As it happens, that's a good thing, because it gives us a larger set of regions, and more subtle patterns of variation, on which to test our ideas. More data is more grist to the scientific mill.

I suspect many of us have an intrinsic feeling for why the tropics have more species than areas closer to the poles. If you've ever tried to keep a plant or animal alive, you'll know that warmth and moisture are important. Cold and dry is generally anathema. We feel the lack of heat and moisture ourselves, for all the beauty of a crisp, dry winter's morning. But warmth and moisture can be good for *individual* plants and animals without necessarily translating into more *species*.

To understand why there are more species in the tropics, we need to go back to themes that we introduced at the start of this book—the fundamental processes of birth, death, immigration, and emigration. Now, though, we need to think about these processes for species.

All life shares a common ancestor, which lived somewhere around four billion years ago. We deduce this from the age of the earliest fossils, and because all life shares the same genetic code. All of the millions of extant species have arisen from this ancestor through the process of speciation—the splitting of a population into different groups of individuals, each with their own characteristics, that can reproduce with each other to produce fertile offspring, but cannot reproduce with members of other groups of individuals.[v] Speciation is how new species are born: a "parent" population splits into two (or more) "daughter" groups (sisters). At the start of the process, all individuals can reproduce with each other.[vi] At the end, individuals can only reproduce with others in each daughter group.

Not all the species that have ever appeared on Earth are still alive today. That's because species can also die—or die out. Giant dinosaurs such as *Tyrannosaurus rex* or *Diplodocus carnegii* ("Dippy") may be the first examples of vanished species that spring to mind, but in fact most species that have ever lived are now extinct. Just like the individuals that constitute them, all species are born and die. Just like us, they survive for some unspecified and uncertain time between these bookends.

---

v. It would be possible to write an entire book on different definitions of species alone, so I'm just going to pick this one and stick to it. It's a fairly standard definition. What it means is that members of one species cannot produce viable offspring with members of another species. This is not always true, but the fact that it isn't is in part because speciation generally takes time, and we encounter "species" at different points on the road to separation.

vi. Given the limitations of sexual compatibility.

Between their birth and death, species can also relocate. The importance of migration again. Individuals moving out from their place of birth to seek new opportunities inevitably take their species identity with them.[vii] Species born in one location can end up living in others—our own species is one example. From the perspective of the numbers of species living in different regions, we generally only consider immigration. That's because it's rare that all members of a species up sticks and relocate. Some ancient humans left Africa, not all of them. We only really need to think about migration adding species to regions, not subtracting species from them.

Remember our equations from chapter 1: the number of individuals living in an area changes because of births minus deaths, plus immigrants minus emigrants. The same is true for the number of *species* living in an area, too: this changes because of births (speciation) minus deaths (extinction), plus immigrants minus emigrants (this last group being rare enough to ignore). Ultimately the fate of the species is determined by the fate of the individuals. The birth, death, and migration of a species needs the birth, death, and migration of the individuals that constitute it. The dance of these processes over the four billion years or so that life has existed on Earth has given us the patterns in species richness that we see today: some parts of the planet home to many species, others home to few. The big question is: What calls the tune for this dance?[viii]

For species numbers to increase, a population needs first to split.

One way this can happen is that it becomes divided by a physical barrier. A mountain range grows, leaving individuals on different slopes.

vii. At first at least. Relocation can start the separation that leads to speciation.

viii. To some extent, the answer to this question depends on whether the number of species in different regions has reached a plateau—termed *equilibrium*, a carrying capacity for species, as we see in logistic population growth—or can still increase— as in exponential growth, termed *nonequilibrium*. I largely work from the assumption that nonequilibrium is probably true. It's hard to be sure when we don't know what the future will look like, though.

Sea levels rise, cutting off peninsulas, making hills into islands. Drying causes forests to retreat, splitting large tracts into isolated woodlands. Over time, differences in the environments occupied by the daughters—or just the blind action of chance as populations drift apart genetically—eventually leads to sets of individuals that cannot interbreed if they meet up again. This is *allopatric* speciation—speciation in "another place." The number of species has increased.

Splitting can also happen without a physical barrier—*sympatric* speciation, speciation in the "same place." A moth that develops on one food plant colonizes a second, and eventually the individuals on different plants diverge enough to be different species. The European Corn Borer is showing us this in real time. Its common name identifies why this moth is considered a pest. But it didn't originally evolve to bore corn—the moth has colonized this crop in the years since people introduced corn (that is, maize) to Europe. Individuals that now develop on corn differ in their sex pheromones and emergence timing from those that develop on other food plants nearby. They are becoming reproductively isolated. The end point of this journey will be two species where there was just one before.

The birth of species increases the number living in an area. Why should there be more speciation in some areas than others? One answer is implicit in the previous two paragraphs—time.

Speciation is a slow process. We've been watching it happen in the Corn Borer, an ongoing separation that has so far occupied the 500 or so years that corn has been grown in Europe. But that's a relatively rapid pace. Humans and Neanderthals are thought to have diverged somewhere in the region of half a million years ago, but were likely to have continued interbreeding until Neanderthal extinction around 40,000 years ago. The Common and Iberian Chiffchaffs are two species of bird so similar that, until a few years back, they were regarded as the same species. They still interbreed from time to time where their populations overlap. Yet, comparison of their DNA identifies evolutionary paths that diverged around two million years ago. A million years of separation might be a typical amount to allow the complete transition into different species. Britain has only been an island for around 1 percent of that time. It's little surprise that we have almost no species unique to our shores.

Because speciation needs time, we expect more of it, and so more species, where more time has passed. So does it follow that the tropics must be older than the temperate zone, given the distribution of species? Well, yes and no. No, in that land (and sea) has existed at higher latitudes as long as it has at the equator. But yes, in that eons of fluctuation in the global climate have had different effects at different latitudes. Those fluctuations have given more time to the tropics.

Our planet is currently in a phase of alternating glacial and interglacial periods that has been cycling for about the last 2.5 million years. At the moment, the cycles take roughly 100,000 years, during which we get warmer interglacials—like now—and colder glacial phases—"Ice Ages," when global temperatures drop and ice spreads equator-wards. It's only been 12,000 years since most of Britain was under a thick layer of glacier. These cycles of glaciation are the most recent expression of a phase of global cooling that started around 34 million years ago. Before that, the planet had experienced a much warmer *greenhouse* period that started around 260 million years ago. Greenhouse periods are the more typical conditions—no continental glaciers, polar regions forested, and tropical climates extending over most of the planet.[ix]

The upshot of all this variation is that the Earth has a much longer history of hot, tropical-style environments. Our current temperate zones have been tropical for much of the history of terrestrial life, and periodically scoured by glaciers for chunks of the rest. In terms of opportunities for speciation, time has not been on their side —unlike the tropics. And this is one likely reason why the tropics have more species.[x]

The ebb and flow of climates across the face of our planet identify another reason why we might expect more species in the tropics. For most of the history of life, there has simply been more tropics. Once again, size might matter. In this case, area.

---

ix. Though equatorial zones may have been too hot for most (terrestrial) life to endure.

x. It seems inevitable that time will matter for species richness, because it takes time for species to form. It's not inevitable, though. It could depend on whether or not species richness has reached equilibrium. Time may be less important if it has, because then more time doesn't necessary add more species. Other processes may matter more.

Area matters for speciation for several reasons. First of all, it allows for larger and more extensive populations. Larger populations are more likely to find subpopulations being separated off, for example by a growing mountain range or a widening river. Larger areas are also, by chance alone, more likely to see the imposition of such barriers. This will give them a wider range of environmental conditions for species to exploit—different elevations, for example.

Larger areas provide more opportunities, but they also improve the chances that species will be able to use them. The larger populations in larger areas will also have greater genetic variability. Variation in the genes translates into variation in the features of the individuals they program, and this increases the chances that some of those individuals will be ready to exploit new environments when they encounter them. This all promotes the birth of new species.

Area matters for the death of species, too. As we have already seen, larger and more extensive populations are more robust against the vagaries of chance. Fire, flood, or just a run of demographic accidents are unlikely to drive widespread and abundant species to extinction, but can certainly do for species that are rare and restricted. Small populations also tend to lack the genetic diversity that would let them adapt when conditions change.

So larger areas are doubly advantaged when it comes to accumulating species. They will tend to have more new species arise, and fewer of those species disappear. Tropical areas are large now, and have been even larger over most of Earth's history.[xi] Larger area and more time equals more species.

This certainly sounds plausible, but hypotheses without data are just stories. As it happens, our best data also support these ideas. Biogeographical regions richer in bird or mammal species are those that have had the combination of a lot of area for a long time. Older and larger regions have had more space and time to accumulate species—and may continue to do so into the future. The geological history of a region matters for how many bird and mammal species live there. Longer and larger gives you more.

---

xi. Beware map projections that distort areas such that the tropics look small—they are not!

A third advantage for tropical life is the availability of energy.

Remember that all of life runs on energy. For most species, the ultimate source of this is the sun—directly for plants and other autotrophs that harvest solar energy to fuel their production of organic molecules, and indirectly for those species that feed on plants or on other animals that plants have fueled. The sun's rays strike our planet more or less perpendicularly at the tropics, but their incident angle increases as one moves closer to the poles. This means that the same amount of energy is spread over progressively larger areas as latitude increases: each square yard in temperate regions receives less overall than in the tropics, with polar regions getting least of all.[xii] For those of us raised in the temperate zones, the ferocity of the noonday sun in the tropics can come as an unpleasant surprise. But that extra energy is a positive boon for life— as long as there's also water, which living organisms need just as much. The luxuriance of growth in the wet tropics reflects the abundance of these essentials.

Given that life depends on energy, it makes sense that areas with more of it ought to be able to support more life. Of course, more life does not necessarily translate into more *species*, but the arguments for the advantages of area apply here too. More energy feeds more individuals. More individuals give more opportunities for speciation, and more resistance to extinction. Energy times area times time equals species. To see the effect of removing energy from the equation, check out the vast continent of Antarctica—lots of area, but little of the sun's energy reaching it.

The sun gives energy to life, but it supplies it to the physical environment, too. Adding energy to the molecules of a solid, liquid, or gas raises its temperature—which is a measure of the thermal energy of a material. Many chemical reactions go faster if you raise their temperature. Up to a point, the physiological processes that underpin life are among them. Hotter temperatures promote higher metabolic rates, which means that species can run through their life cycle more

xii. The exact distribution varies between winter and summer in each hemisphere due to the Earth being tilted on its axis relative to the sun.

quickly. Evolutionary change happens via reproduction,[xiii] and so the less time between generations, the faster that speciation can occur. Mutation rates may also be higher at hotter temperatures—mutations provide new genetic variation for evolution to work with as populations diverge. So another way in which tropical climates might end up with more species is that the processes that drive diversification motor along faster there. At least for species that depend on the environment for their body temperature, like insects. Warm-blooded vertebrates like you and me maintain a constant temperature regardless. This mechanism shouldn't apply to us.

Again, these hypotheses sound plausible. Again, we need to see what the data say. Again, our best data say that energy matters.

After area and time, energy matters most for bird and mammal species numbers. All else being equal, warmer is richer. That's not because diversification goes faster in warmer regions, though. The bird and mammal family trees don't provide any evidence that the *rate* of diversification matters much to variation in the richness of regions. But then we didn't expect temperature to speed diversification rate in warm-blooded animals. Yet it's also not because warmer areas have more food, because temperature matters more than plant productivity. So why *does* temperature matter for birds and mammals? The answer is probably quite prosaic. Life needs water in its liquid form, and it's harder for species to cope in parts of the world where it freezes. So fewer do.

Energy also matters for ectothermic species like hawk-moths. Most hawk-moth species like it hot. But hawk-moths flip the situation we see in birds and mammals: plant productivity does a much better job than temperature in explaining their species richness. Energy matters, but through the availability of food resources, not by enabling these ectothermic insects to keep warm.[xiv] Simply, more food equals more moths—giving more opportunities for speciation, and more resistance to extinction.

xiii. This is why we know the egg came before the chicken. The chicken is genetically identical to the egg that it hatched from, but the egg is genetically different from the chicken that laid it.

xiv. Or by speeding their evolutionary rates.

It's not only the amount of resources that may matter for species numbers, but also their reliability—whether boom is followed by bust.

When the world is stable and consistent, species have leisure to adapt to, and specialize on, the use of a specific resource—be that a habitat, or a food source within that habitat. They don't need to be able to tolerate a wide range of environmental conditions when the climate is predictably clement. They don't need to hedge bets with a broad diet when their food is always there. Specialization likely promotes speciation—as with the European Corn Borer, for example. Extinction is also less likely in a reliable environment—the vagaries of chance are less likely to bite, even for populations that are small and restricted in extent.

Contrast that with the situation facing species in variable and unpredictable regions. They need to be able to cope with a wider range of conditions thanks to dramatic changes between the seasons—summer and winter pose very different challenges in the temperate and boreal zones compared to the tropics.[xv] Let alone that they might see four seasons in one day. Specialization is not so easy if the resources on which you rely might suddenly disappear. Variability brings chance into play, and small populations of highly specialized species will be more likely to suffer. They don't have the cushion of fallback options. Extinction will tend to weed them out, leaving communities of mainly widespread and catholic species. Lower rates of speciation and higher rates of extinction will lead to fewer species in unstable regions.

These differences may also explain why the floras and faunas of tropical and temperate zones are largely made up of different species. Few, even our own, are equally at home in both.[xvi] Those raised in stable tropical conditions are generally ill-equipped to cross the boundary into harsher seasonal environments. Those that do need special adaptations to survive—layers of fat, heavy coats, cells laced with antifreeze—requiring the expenditure of precious energy that could otherwise be

---

xv. The tropics are seasonal, but the changes and extremes tend to be less dramatic—especially in temperature, which as we've just seen is critical.

xvi. We've done a good job of finding ways to make ourselves at home, though, from clothes and heating to (increasingly) air-conditioning.

spent on reproduction or growth. Temperate species attempting the reverse journey are likely to lose out in competition in the tropics if they don't quickly shed those expensive adaptations. Trade-offs again—a species can't be good at everything. So, on average, it's rare that a species steps across the regional boundary. The implication is that the birth and death of species matter more for species richness than does migration.

That said, many groups of organisms *have* originated in one region and moved to another—usually heading out of the tropics. Flowering plants, amphibians, and birds are among them. Crossing a regional boundary does not have to be a common event to matter, if it opens up new opportunities for the pioneers to diversify and speciate. Just how great those opportunities are is likely to depend, again, on time, area, energy, temperature, and stability in the region colonized—the same features of the environment that favor the tropics as the source of immigrant species, through their influence on the chances that species are born and die, and the weight of numbers that results.

These effects can explain why tropical mountain ranges are notably rich in species, such as hawk-moths. Species may not often step across latitudinal boundaries, but the challenges are reduced when those steps are altitudinal—a few hundred yards up a tropical mountain slope. Species don't even have to move, just ride the uplift as mountains are formed, diversifying as they go. The average climate changes with elevation, from lowland tropical forest through to high altitude grasslands and tundras, but these climates are also stable. They don't have the same seasonal extremes that occur when changes are with latitude. So mountain slopes in the tropics provide a high diversity of stable environments, not only giving species new opportunities, but also leisure to adapt to them. Reliability matters.[xvii]

So far, all of the factors I've discussed that might promote species richness have been features of the physical environment. What about

---

xvii. As it happens, the habitats on tropical mountains richest in hawk-moths are also those with the largest land areas. Size matters on the slopes, too.

the interactions between species that preoccupied the early parts of this book? Can they also affect species richness? Perhaps.

It's possible that interactions between species matter more in tropical regions, because there, vagaries of the physical environment matter less. If damage or death mainly comes from other species, that will spur the loser to evolve to evade the winner. In competition, that means finding ways to live where their opposition is weak—moving apart in niche space. Recall the many defenses moths have developed to avoid being eaten, which all speak to the evolutionary pressure of interactions. The need to counterattack then drives evolution in the consumers. The result might be a rapid evolution of interacting species—and more species if different populations are driven in different directions. Cooperation between different species—pollinators and flowers, the partners in lichens, for example—can have the same effect. Interactions between species may speed the rate at which speciation occurs.

It's also possible that interactions can reduce the likelihood of extinction. Seedlings that sprout in the shadow of their parent are likely to suffer more from herbivores and pathogens. Those consumers are probably already on hand, attracted to the parent—a strong imperative to flee the nest. This gives an advantage to offspring that disperse farther, but also to rare species. Their seedlings are less likely to disperse and find themselves near another relative. An advantage to being rare prevents species from dominating communities, and driving competitors to extinction. Remember the processes stabilizing communities: species coexist when interactions within species matter more than those between. In this case, the interactions are mediated through consumers. Knock-on effects on herbivores (and their predators), or rare species effects in consumers, can propagate these effects up the food chain. More species is the result.

Interactions certainly matter. Competition and predation put the brakes on population growth, and drive adaptation within those populations to evade their impacts. But all species interact with others. Biotic interactions may be more intense in the tropics, but do they *cause* higher species numbers there, or are they a *consequence* of those numbers? At best, interactions probably only build on advantages provided first and foremost by the physical environment.

Our megafauna suggest that this is indeed the case. We have little evidence that bird or mammal species are further apart (or indeed closer together) in niche space, or in their morphology, in more species-rich regions. Yet we'd expect such differences if interactions are generating diversity. Interactions undoubtedly influence the coexistence of species in local communities, and competition certainly drives related species apart. But when it comes to how *many* species coexist together in broad biogeographical regions, the physical environment is king.

Interactions are not irrelevant, though, as we will see.

Variation in the processes that drive diversification not only lead to variation in how species richness accrues over space, but also over time. The moth trap gives us a window into this, too, through what diversification does to the number of species in different groups of moths.

Recall the identification challenges faced by my students at Blencathra. Finding a Poplar Hawk perched on the outside of the first trap was certainly a boon in hooking them into the joys of moths, but this species is hardly typical of the catch. Most moths are something else, even if we're only trying to identify the larger, easier macros—mostly either noctuids (owlet moths) or geometrids (loopers). Three quarters of the species the students caught (and 71 percent of the individual moths) were from these two macromoth families—even though another thirteen families have British representatives. When it comes to species numbers, families are not all created equal, either. There are very much haves and have nots here, too.

Just as all life shares a common ancestor, so too do all moths. A few wing scales dating from around two hundred million years ago are the earliest evidence of them in the fossil record, but the oldest fossil is never the first of its kind. The genetic distance between different groups implies that the true origin of the Lepidoptera could have been as far back as 300 million years ago, in the late Carboniferous or early Permian geological periods. This would be when they first diverged from their closest insect relatives, the Trichoptera (caddis flies). The planet then was in a cool climate phase and so those unknown first moths could have been tropical or temperate. Either way, all of the stupendous diversity of moths alive today—160,000 known species, including those that convention calls butterflies—is descended from that single ancient ancestor.

That we see so many species of moth is probably down to a big stroke of luck—the evolution and diversification of the angiosperms (flowering plants), which blossomed at around the same time as moths started their evolutionary expansion. The larvae of the earliest moths are thought most likely to have fed on detritus or moss, and would have grown into adults with chewing mouthparts. The shift onto angiosperms was probably made by species that developed inside plant structures, perhaps as leaf miners. Moths then piggybacked on the success of the plants.

More than that, moths may have helped plants succeed, thanks to their contributions to plant sex. The next big step for moths was the evolution of the proboscis. This may have developed to drink water or sap, but allowed them also to exploit the energy provided by nectar. Plants exploited the moths back by dabbing them with pollen.

Larvae shifting to develop *on* plants, rather than within them, would have been freed up to grow larger. The "true" macromoths, comprising the noctuids, geometrids, hawk-moths, and their relatives (hook-tips, lappets, emperors, and silk moths) probably shared a common ancestor around ninety million years ago.[xviii] Owlet moths may have started to diversify about fifteen million years after that, before the extinction of the dinosaurs,[xix] although their major proliferation probably happened after the asteroid strike, sixty-six million years ago, that sounded the death knell for that group. Hawk-moths are relative newcomers, appearing only about forty-five million years ago.

The moth family tree has many branches, but different branches have had very different degrees of success, judging by the number of descendants they have produced.

---

xviii. Some of the families that by convention we include in the macromoths are technically large micromoths, and so not "true" macros. The Cossidae, for example, which includes the GOAT. Yes—the Goat is really a micro!

xix. Non-avian dinosaurs, I should say. Birds evolved from theropod dinosaurs, and so technically are dinosaurs, too.

The first fork in the tree probably led to the superfamily Micropterigoidea down one branch, and all other moth species down the other. There are around 160 known species of Micropterigoidea, versus (more or less) 1,000 times as many other Lepidoptera—even though both branches have had exactly the same amount of time to diversify.

Asymmetries in the net outcome of the births and deaths of species are common. Around ninety million years ago, the common ancestor of all true macromoths split into two species. One lineage diversified into the 700 or so extant species of hook-tip, like the Barred and Scalloped species that graced my Lewtrenchard trap. The other gave rise to one hundred times as many species—roughly 42,000 noctuids, 23,500 geometrids, 2,000 lappets, and 1,450 hawk-moths. Haves and have nots—just as with different regions around the world. When it comes to understanding the composition of a moth trap, understanding why there are so many more owlets and loopers than hawk-moths is another part of the puzzle. Why *do* groups differ so much in how many descendant species they have spawned?

In broad terms, the answers are similar to those for the richness of regions. Opportunity is key. One element of this is again that provided by time.

It's inevitably going to be true that a taxon that has been around for longer is likely to have more species. In most cases, that's simply because daughters are younger than their parents. There are more species in the noctuid family than in the plusiine subfamily (Silver Ys and their relatives) because the latter is a younger subset of the former. The effect of time is not always so straightforward, though. Noctuids are both older and more species-rich than hawk-moths. Time may not be the only reason for this difference, but diversification takes time, so it helps.

That said, it can't all be about time. As each other's closest relatives, Lepidoptera and Trichoptera are the same age. Yet there are ten times as many moth as caddis fly species. See also hook-tips versus owlets. Other explanations are needed.

One of these is area, again. Any group restricted to a small region will not be able to attain as high a diversity as that in a larger one. There will be fewer opportunities for species to split. Species that do appear will have smaller geographic ranges, and so be more prone to extinction. Birth rates for species will tend to be lower in small areas, and death rates higher. We expect fewer species to accumulate as a result.

This is illustrated by a classic example of diversification in moths: the genus *Hyposmocoma*. This is a group of more than 600 species (with many more probably awaiting discovery), every one of which is found only on islands in the Hawaiian archipelago. That's one in every three species known from its family (Cosmopterigidae) worldwide, and a similar proportion of all moth species known from the archipelago. In contrast, these islands are home to just two native butterfly species.

Hawaii is a fascinating place to study diversification. The chain of islands that comprise the archipelago was formed as the Pacific tectonic plate moved over a hotspot of volcanic activity. Islands grew as lava from the hotspot bubbled up above the ocean's surface—a process we see ongoing today on the largest and youngest, Hawai'i, also known as the Big Island. Each island in the chain had a brief moment (in geological time) of growth, which ceased when the plate moved it away from the source of the lava. Erosion by the waters of the Pacific Ocean has since caused islands to wear away. The oldest ones are now also the smallest.[xx] The hotspot has been active for perhaps eighty million years.

It's thought that the ancestor of today's *Hyposmocoma* moths arrived from the northern boreal zone around fifteen million years ago, before any of the current large islands were even born. This again points to the importance of migrants—the ultimate origin of all Hawaiian (terrestrial) biodiversity. They speciated on the islands available back then, but faced catastrophe: homes that were slowly eroding away from under them. Fortunately, new potential homes were also appearing, as the volcanic hotspot continued to make new islands. The moths migrated again, this time along the island chain, evolving into new forms as they went. These include the only moths whose caterpillars are known to predate

---

xx. The very oldest of the Hawaiian Islands are now so worn away that they do not even break the ocean surface, but are still present as submarine seamounts.

upon snails (snails being another group that diversified dramatically after migrants reached the archipelago).

Today, the four largest islands in the Hawaiian chain all have very similar numbers of known *Hyposmocoma* species. The Big Island, still growing, is only around 400,000 years old. It's had much less time for moths to diversify than the three million years or so available to Oahu. *Hyposmocoma* must have arrived on the Big Island more recently, giving it less time to diversify into multiple species. Where the Big Island wins is that it's, well, big—at around 4,000 square miles, roughly seven times Oahu's size. Area compensates for youth when it comes to accumulating moth species. But both age *and* area matter for diversity—over space *and* over time.

Asymmetries in diversification also come from the opportunities provided by the abundance of resources. Again, more resources are likely to translate into more species. The energy-rich tropics are home to more species, and so clades that have diversified there are generally also more species-rich than those based in the temperate zone.[xxi] In the hawk-moths, for example, the subgroup Dilophonotini is based in temperate Asia, and has only twenty-five species. Its sister is the tropical Macroglossini, which has had the same time for divergence to occur, but with 502 species the result. Resource abundance may well be responsible for the difference. We already know it promotes hawk-moth diversity over space.

When it comes to resources and opportunities for moths (and many other animal groups), plants are key. Moth and plant family trees suggest that important periods of diversification in both seem to have coincided—implying the possibility of cause and effect, in one or both directions. One spurt in moth diversification happened around a hundred and fifty million years ago, alongside the first divergences in the Mesangiospermae, the group that includes almost all extant flowering plant species. The growth in plant diversity would have given opportunities for moths to colonize new food species, once evolution had worked out for them how to sidestep the chemical and other

---

xxi. A clade is a group of organisms, technically all the evolutionary descendants from a single common ancestor.

defenses that plants deploy. Diversification was also aided by switching feeding location on the same host plant—moving from stems to leaves or flowers, for example—and switching feeding mode—such as from mining to chewing leaves. Finding a new way of life, and doing it better than other species.

The most spectacular burst in moth diversity, though, was around the time of the Cretaceous–Tertiary boundary, near the demise of the dinosaurs. It concerned the owlet moths and their close relatives. The diversification rate in this group increased to more than seven times the background level in Lepidoptera—a massive acceleration in the speed at which new species arose. These moths were clearly taking advantage of an unusual opportunity, and it was probably provided by a parallel burst of activity in herbaceous plants—those species that, as we know, caterpillars of many British moths like to eat "a range of."

The ancestral condition for owlet moths probably involved developing on woody plants, judging by the habits of the species around today. For example, Notodontidae (Prominents) is one of the oldest owlet families. It's represented in the UK by such species as the remarkably camouflaged Buff-tip, whose caterpillars feed on leaves of a wide range of deciduous trees, and the Puss Moth, that ermine-clad beauty whose caterpillars feed mainly on poplars and willows (and have startling false faces to scare off predators). These are temperate examples, but species in the ancestral owlet moth clades are overwhelmingly tropical in distribution. Their early diversification coincided with a period of high global temperatures, including the Paleocene–Eocene thermal maximum around 55 million years ago, when atmospheric carbon and global temperatures briefly spiked. In a time of extensive tropical forests, and even forests in the polar regions, tree-feeding species made hay.

Through the Eocene, temperatures progressively cooled, to the point that polar ice caps started to form. The climate also became drier. Forests retreated through this period, thinning out and becoming patchier. Herbaceous plants moved into the gaps. Temperate climates enlarged as tropical ones shrank. As we've already discussed, it's hard for species to cross the boundary from tropical into temperate regions. But the benefits of success are abundant new opportunities for any that can make the jump.

Owlet moths are a group that seized those opportunities with all six legs. Species from younger taxonomic groups within the clade are temperate and tropical in equal measure. Most of the temperate species feed on the herbaceous plants that flourished in the more open habitats there. Those crossing into temperate zones have had to modify their life histories to cope. They allocate resources differently over their lifespans to allow diapause,[xxii] and have developed mechanisms for cold tolerance. Polyphagy, or the ability to eat a variety of food, develops often in the temperate zone, as many food sources are less stable and predictable over space and time. The ability to move to track resources matters, too. A prime example of a polyphagous and dispersive owlet moth is the Silver Y we met in the previous chapter. It's a member of a subgroup that has radiated widely, and almost all of its species are temperate.

The temperate zones have provided opportunities that owlet moths have grasped. That's why their species are the most numerous of the macromoths I catch in my temperate moth traps. That said, they still haven't diversified faster overall in temperate regions than in the tropics—the tropics still hold sway in terms of total species number, for all the reasons we've already seen. But success in the tropical *and* temperate zones has made Noctuoidea the richest of all the Lepidopteran superfamilies. There are now around two species of owlet moth for every species of butterfly worldwide. Butterflies are show-offs that get lots of attention, but the under-the-radar owlets are no less interesting, and deserve just as much.

So much in life comes down to luck.

It was lucky that Kindrogan had the moth trap that inspired me to take up mothing. It was lucky that I had a successful academic career that put me in charge of a field course. It was lucky that I had parents who supported me in my education and choices. When we construct the narratives of our lives, it's easy to be seduced into thinking that

---

xxii. Periods of suspended animation, allowing individuals to hunker down when conditions are unfavorable.

our successes are down to our talents, innate or acquired through hard work. Never forget the influence of blind chance. You are just lucky to be alive.

Luck equally applies to the rest of the natural world. When we open a moth trap and peer inside, we are seeing the outcome of processes that weave the deterministic with the stochastic. Rules and luck together. The number of species we catch is a perfect illustration.

The British Isles are home to around 2,500 species of moth, but count among the poorer areas of the world in this regard. Why we have this many, and not 1,500 or 3,500, we cannot say with any precision. What we *can* say is that if you want richness in your moth trap, you need to move closer to the equator.

It's not inevitable that more species should live in the tropics, though.

Evolution needs material to make new species, and it needs time to work. Larger areas can house larger populations, providing the raw material for time to turn into species. It's not inevitable that the tropics are larger, though. It's not inevitable that time has been on their side. The movements of tectonic plates and the procession of land across Earth's surface, the rise and fall of carbon dioxide in its atmosphere over millions of years, and the wobble of the planet on its axis, have all conspired to favor terrestrial life at low latitudes.

Life likes it hot and wet, but within limits—the Goldilocks zone. If it's *too* hot, plants and animals cannot cope.[xxiii] This may well have been the case during greenhouse periods in Earth's history. The molecules that make up life denature, with potentially fatal consequences. Right now, though, the tropics are just right. That accentuates their advantage in the raw material for speciation. Luxuriant plant growth provides lots of opportunities for consumers, and consumers of consumers. Especially when that growth is reliable, as it generally is in the tropics.

The dance between speciation and extinction, and how species redistribute themselves between their births and deaths, has given us dramatically different outcomes in different parts of the world. The numbers have very much come up for the tropics, thanks to a combination of good luck and demography. Yet not all groups have

---

xxiii. Or too wet, for life on land. It can never be too wet for aquatic species.

played those numbers with the same degree of success. Again, chance has had a hand.

Perhaps 300 million years ago an insect population split in two. One sister became the founder of the caddis flies. The other gave rise to the moth dynasty—one of the most successful branches of the tree of life, accounting for one in every nine known species. But we know how easy it is to lose populations through the vagaries of chance. How lucky did that first moth have to be? What other remarkable forms might we have lost because the odds didn't fall in their favor?

Moths have been successful, but some have been more successful than others. Those that moved to exploit early angiosperms had no inkling that they were backing the right horse, but they have won big. Groups that have benefitted from the effects of time, area, energy, and stability in the tropics have also done better, but some that took the opportunities to be gained from crossing the boundary to temperate latitudes have benefitted from this gamble. It's likely that being the first into such habitats has reinforced this success. The same priority effects that influence the composition of local communities also reinforce differences in diversification. There would have been numerous other gambles that didn't pay off, but then history is written by the winners. Ecology is written to explain why they won.

It's inevitable that most of the macromoth species in the Blencathra traps that week in June were owlets or loopers. Most species in the UK are from these two groups. But it was lucky that that first trap was adorned by a member of one of the less diverse families: that glorious Poplar Hawk-moth. I know that it inspired the students to think more about moths. I hope it also inspires them to consider how we can make the world a better place for moths and their ilk. Goodness knows they need our help—as we're about to see.

Chapter 8

The Box-tree and the Stout Dart
How Ecology Is Now Humanity

Man can hardly even recognize the devils of his own creation.
— Albert Schweitzer

Box-tree Moth, Camden, London.

197

A roof terrace in urban London seems like an unpromising location to be inspired by nature's riches, and in isolation it would be. But as we've seen again and again, all of life is connected. From the mature trees and well-tended gardens that the terrace overlooks, through the wider network of gardens and parks in London, to the forests, meadows, pastures, and heaths that encircle the city, the country and continent beyond, and continents beyond that, processes underpinned by the fundamental rules of nature work to bring life to my terrace. They bring my terrace to life. Every time I open my moth trap, I am looking at the culmination of a four-billion-year-long story, whose action spans the globe. Every insect inside has an unbroken connection, parent to offspring, that goes back to the very first organism that appeared on our planet. Each one has a role in the greatest story ever told—the story of life. On the only planet in the universe known to house it.

The most recent chapter of this story has revealed a worrying plot twist, however. The first moth I tried to identify, that first morning on my roof terrace in London, could not have been better chosen to make the point.

As a moth-trapping *ingénu*, it made sense for me to start with the easiest species. There was an obvious candidate. It was one of the largest moths in the box, and with distinctive markings. None of the many subtle shades of gray or brown, but bright, pearlescent wings edged thickly around with black. This printed on a distinctive outline, with slightly concave leading forewing edges curving round to a subtly pointed tip. Pale "socks" on each leg, and long antennae laid back and reaching almost to the tip of its snow-white abdomen. I had a strong feeling that it was a species I'd seen before, although I couldn't pull its name from my memory. Not to worry, though—a couple of minutes with the field guide would sort it out.

Twenty minutes later, and I was ready to cry. I'd been through the book twice, and nothing illustrated there fitted my moth. The closest was the Clouded Border, another species with black-edged white wings. But the resemblance was superficial at best. On top of that, Richard Lewington's brilliant field guide illustrations are (with a few exceptions) reproduced life-size. I could place my moth next to his painting and see immediately that it was too large. Also, the wrong shape. And the wrong pattern in the monochrome. How was it possible that nothing matched? I had a trap full of moths to identify, and I'd fallen at the first hurdle. What hubris, thinking I could casually pick up a new set of animals to master. I was on the verge of packing up mothing for good.

At this point, I reached for one of the key tools in the scientist's kit— one that can be dangerous in the wrong hands, but that's phenomenally useful if you learn how to use it properly. Google.[i] I typed in "black white moth UK" and clicked Search. A second later, and there was a picture of my animal. I had my identification: Box-tree Moth.

The identification explained why I hadn't been able to find my moth in the field guide. The guide I was using only dealt with macromoths. Despite its size, Box-tree is a micro. In this case, though, my micromoth field guide would have been equally useless—that didn't depict the Box-tree, either. Simply, the moth had not colonized the UK when my guides were printed. That also explained why I had a vague memory of the species: it's an alien.

The Box-tree Moth is native to China and Korea, so when some adults were discovered in Baden-Württemberg, western Germany, in 2006, they were far outside the species' normal purlieu. We've already seen that some moths are capable of impressive feats of long-distance migration, but this journey would have been beyond even travelers like the Silver Y. Unaided, at least. But the Box-tree Moth didn't make that journey under its own steam—it hitchhiked.

As I'm sure you'd have predicted, Box-tree Moth caterpillars develop on the leaves and shoots of Box plants (species in the genus *Buxus*). Box is common in the horticultural trade. Its glossy, evergreen leaves and compact growth form make it ideal for hedging. Box plants are regularly exported from East Asia to Western Europe, and no one is going to check every plant for unscheduled passengers. Or spot every one if they did. It was only a matter of time before Box-tree Moths (egg, caterpillar, or pupa) caught a ride. Their arrival in Germany—and then their appearance the following year in a commercial polytunnel in Kent, in southeast England—was actually not that much of a surprise.

---

i. Plenty of other search engines are also available.

Like the Gypsy Moths in Massachusetts, the Box-tree Moth in England initially spread quite slowly. By 2010 there had still only been ten records of the moth here, and the annual number of records only topped one hundred in 2015. As in New England, though, this was just the calm before the storm. A drastic increase in numbers was noted in 2018, when I caught my first one. More than 9,000 Box-tree Moth records were logged in England that year, and the species had already spread to Wales and Scotland. Once again, the worms were very much out of the can.

Aliens are one of the main ways in which humans are changing the patterns of life on this planet. In the ecological context, they are species that do not naturally occur in a place, but have been moved there—beyond the boundaries of their native distributions—by human activities. They are also the subject of most of my scientific research, which was why that first Box-tree Moth I caught rang a bell. Not that it's possible to remember every species that's alien somewhere—by 2005 there were already more than 16,000 of them worldwide, including hundreds of species of moth, and numbers continue to grow rapidly. As I ran more traps, I encountered more examples. The Light Brown Apple Moth, originally from Australia, that despite its name eats almost any greenery. The Ruddy Streak, another Antipodean, a consumer of leaf litter. The Oak Processionary Moth from central and southern Europe, whose caterpillars possess hairs that can cause severe skin irritation, and respiratory problems if inhaled. All originally arrived on plants imported to the UK from elsewhere in the world. All now come to my roof terrace.

Aliens worry scientists because they can have substantial impacts. We've already seen what happened in Medford, and while less dramatic in its numbers, the Box-tree Moth also gives us cause for concern. It's potentially a problem for owners of Box hedges, as the caterpillars are capable of chewing plants down to their stems. That's an inconvenience for sure, but other hedging plants are available. More worrying is the potential for the moth to destroy wild Box woodland, which is a scarce habitat in Britain. In 2009 and 2010, Box-tree Moths ate their way through 90 percent of the foliage in Germany's largest Box forest, and within a couple of years those trees that had been completely defoliated were dead—more than a quarter of the total. It's changing the very nature of this habitat. In 2018, caterpillars were found at the famous English site for *Buxus*—Box Hill in Surrey. The future for Box there looks bleak.

The growth in the numbers of aliens means individuals, populations, and species are being added to floras and faunas around the world. Unfortunately, they are also being subtracted. Again, there is no better illustration of this than the animals that turn up—or that don't—in my moth trap.[ii]

The Beaded Chestnut is one of the classic moths of autumn. I generally catch a handful on our trips to Devon over October half-term, if the rain relents. Like many species at this time of year, it's dressed in the colors of a dead leaf—crosslines, ovals, and dots in various shades of brown, set on a gingery background, that gives it its English name. Its caterpillars feed on—for variety—"various herbaceous plants," and in a wide range of habitats. There ought to be no shortage of opportunities for the Beaded Chestnut in Britain, but its population there is a shadow of its former self: its numbers declined by 92 percent between the years 1970 and 2016. It does still occur in London, but it's never appeared on my roof terrace.

Another species named for its hue is the Orange Moth, a large Geometrid, vividly colored to a degree that puts many butterflies to shame. It's on the wing in June and July, and I was excited to catch one during our lockdown residency in Devon in 2020. Lucky too, because Orange Moth disappeared from more than three-quarters of its British distribution between 1970 and 2016. It's unlikely now to appear in my London trap, while only time will tell whether it continues to hang on in Devon.

At least I've caught these species. There are many that I haven't, and

ii. Figures in this section come from a range of publications, including the *Atlas of Britain & Ireland's Larger Moths*, *The State of Britain's Larger Moths 2021*, and *The Changing Moth and Butterfly Fauna of Britain during the Twentieth Century*. Full citations are given in the list of sources.

for which the odds that I ever will are getting longer every year. The Lappet is a large, purplish-brown fur-ball of a moth, with scalloped wing margins like a cluster of dried oak leaves.[iii] It's one I lust after catching, but its numbers declined by 97 percent, and its British distribution by almost two-thirds, from 1970 to 2016. The Garden Dart sounds like the kind of species that ought to turn up in my moth trap. Like the Beaded Chestnut, its caterpillars feed on herbaceous plants in a wide range of habitats. Like the Beaded Chestnut, it's in severe decline—numbers have dropped by 99 percent, and it's gone from 85 percent of its British distribution.

Things are worse for the Stout Dart. It's an undistinguished noctuid, described in the field guide as drab and mousy. Once widespread across England, a massive decline in the last forty years has resulted in no records since 2007—a 100 percent decrease in its abundance and, therefore, also in its distribution. It's always difficult to know when the last individual of a population or species has died, but the Stout Dart is probably extinct in Britain. We search our traps for that species now in vain.

It's not alone. Around fifty other moth species have disappeared from the British fauna since 1900. Species like the Cudweed and the Lesser Belle. The Orange Upperwing and Bordered Gothic. A set of species that the odds of me catching are now very long indeed.

Species are being lost from floras and faunas around the world, but is that really a big deal? You know by now that we expect to see turnover in communities. We expect species to go extinct from areas, and we also expect species to colonize. Why should the British moth fauna be any different?

It isn't. In fact, more species have colonized Britain than have gone extinct here in recent decades. We've lost 51 species of moth since 1900, but we've gained 137. Some of those new moths are aliens, like the Box-tree, but many have arrived under their own steam. The most numerous species I attracted on that first morning in Camden was the Tree-lichen Beauty. Thirteen individuals of this jade-cloaked noctuid sat in and around the trap. Yet prior to 1991, the field guide told me, only three

---

iii. Hence its scientific name, *Gastropacha quercifolia*.

individuals had been recorded in the whole of the UK, and the most recent of those in 1873. For a second I thought I'd caught something really rare, before reading on to find that the species is already abundant and widespread in the London area. We've met the Vine's Rustic—another of the commoner species I catch in Camden, and another that was only a rare immigrant to the UK until the twentieth century. The first British record of the Oak Rustic was in 1999. I caught my first in Camden in November 2020.

So the number of moth species in the UK has actually *increased* over the last century. This is surely good news for moth-ers and biodiversity alike, especially at a time when the newspapers feature frequent stories that we face the possibility of insectaggedon—an ongoing catastrophic decline in insect numbers that threatens the collapse of nature. Well, once again, yes and no.

Yes, because species number is a key metric of biodiversity, and generally we think of more as better. That's because ecological communities with more species seem like they should be more resilient to environmental upheavals. Times that are bad for some species are less likely to be bad for all when there are more species in total. When some species are suffering, others can step into the breach to ensure that vital ecological roles are fulfilled. Bad years for some herbaceous plants can be compensated by better years for others. Caterpillars that feed on a range of them can still find food. Declines in bees would mean many flowers going unpollinated, if moths weren't also attracted to the same nectar. We call this *functional redundancy*—more species give us backup capacity when systems are stressed. So more species ought to be good.

As it goes, the evidence that ecological communities with more species are actually more resilient to stress is mixed. As we saw in the last chapter, stability can favor the development of highly species-rich areas, but areas rich in species unused to stress. Introducing new pressures on those species may simply result in the extinction of some, without compensation in the performance of others. This is one reason why aliens can then move into areas with diverse native biotas—if they're species better used to coping with these stresses. Even if that weren't the case, there are still reasons why an increase in the number of British moth species numbers is not an unalloyed positive.

And now we get to our "no." Species number is only one measure of the state of biodiversity. Other measures are looking distinctly less rosy.

Amateur enthusiasts have been catching and recording moths in Britain for hundreds of years, and the National Moth Recording Scheme collates and analyzes this information. NMRS records are *ad hoc*, but we also have a systematic survey of British moths, thanks to the nationwide network of light traps run every night as part of the Rothamsted Insect Survey (RIS). These traps tell us how the populations of more than 400 species of our larger moths have changed over the last fifty years or so. The RIS is the best window onto insect population trends anywhere in the world. The view from that window is disturbing.

Numbers of individuals of these larger British moth species have fallen by 33 percent overall since the late 1960s. To put it another way, the RIS light traps now only catch sixty-seven moths for every hundred they caught when I was a baby. Across the country, that's a loss of billions. Numbers fluctuate—some years are better than others—but the overall trend is one of long-term loss. Declines in the south of Britain have been almost twice as bad as those in the north (39 percent vs 22 percent). And it's not as if the 1960s was a high-water mark in British moth populations. The original Rothamsted moth trap run since the 1930s (by C. B. Williams and others) shows a 71 percent drop in numbers caught before 1950, compared to the 1960s and '70s. Those hundred moths when I was a baby would have been 300 or more when my father was in nappies. We probably shouldn't read too much into just one trap, but if you think you remember there being more insects on the wing in your youth, your grandparents would have thought the same thing at your age. This loss of experience of the natural world with the loss of each previous generation gives us *shifting baseline syndrome*—we think our experience is typical, but we just don't know what the world was like one hundred years ago. That's why data matter. That's why initiatives like the RIS are critical.

The total loss of moths is also reflected in population trends for individual species. Numbers in the traps have decreased for more than two-thirds of species, with about a third showing increases. As we saw in the early chapters, population sizes naturally fluctuate for all sorts of reasons. For some species, we might record a decrease because we are comparing the top of a population cycle in 1968 with the bottom (or middle) of the cycle in 2017 (and vice versa to record an increase). The data suggest that about half of the species have changed in abundance more than we expect from natural variation alone—these are the moths for which we can be confident that changes in numbers in Britain are real. Among these species, four times as many have declined as have increased.

These data are the highest quality we have on insect declines, but we still need to add a caveat or two to our conclusions.

One is that we only have enough data to calculate population changes for common moth species. We don't know in detail what's happening to our rare species. The general tendency for more species to have colonized Britain in the last century than have gone extinct might be cause for optimism here—there are clearly once-rare species on the up. On the other hand, the fact that we are seeing evidence of declines in the common species is much more of a concern. Most moths are members of common species. A 33 percent decline in the billion-strong population of a common moth is not offset by a 33 percent increase in a rarer species with only a million individuals. A thousand such increases would be needed to balance the ledger. Of course, increases can be greater than 100 percent, while decreases cannot. But losses in common species—that historically have done well in their environment—is a real worry.

Another caveat is that while British moths are more likely to be declining in abundance, they are also more likely to be increasing in their distribution. It's a paradox that species are, on average, losing individuals, but being recorded from more locations—a 33 percent loss of individuals, but a 9 percent increase in areas occupied. This is perfectly possible, of course. Species can decline a lot in some areas without disappearing completely, while simultaneously appearing in new areas in small numbers. It's just a bit unexpected. And to add to the confusion, there's debate about whether moth biomass in RIS traps has changed in the same period, although the most comprehensive recent analysis of moth populations does suggest worrying declines. The overall picture is one of fewer moths of most species, spread more thinly across the country.

The UK is just a small archipelago, less than 0.2 percent of the ice-free land area of our planet. Are the worrying declines in moths we see there repeated over the other 99.8 percent? To be honest, we have very little clue.

The picture in Continental Europe seems to be similar to that in those young islands off its northwestern shore. Numbers of macromoths overall are declining in the Netherlands. The same is true in Finland, though numbers of species are increasing, as in the UK. Traps across Hungary show declines in the total number of species caught, but not in the numbers of species or individuals in any given trap. Thirty years of data from Norway show declines in abundance and species numbers, but that's from just one trap. Studies in the Missouri Ozarks, Arizona mountains, and Ecuadorian cloud forest show little evidence of long-term (well, twenty-year) changes in moth caterpillar numbers. Costa Rican caterpillars have declined in numbers during this century, though, with losses from every moth group studied. But it's hard to draw strong conclusions about all of the Americas from just four small-scale studies. From most of the world, we just lack data.

If all we knew about the state of nature came from moths, we might be worried, but not (outside of Northern Europe) with any great conviction. As it happens, we do know about more than moths.

Scientists and others have been recording the sizes of at least some animal and plant populations worldwide for decades now. The best studied are vertebrates—our charismatic megafauna. We have data for changes in more than 20,000 vertebrate populations, across more than 4,300 species, from birds in European woodlands to salmon in Alaskan rivers. Pulling these studies together allows us to check the overall health of these populations since a starting point of 1970—as laid out in the biennial Living Planet Report. This health check suggests a patient that is rather unwell.

The most recent Living Planet Report identifies an average 68 percent decline in the size of vertebrate populations worldwide. Not all parts of the world are losing at the same rate. Eurasia is not doing as badly as the rest, with "only" a 24 percent decline on average. Likewise North America, with a 33 percent drop.[iv] Africa comes in at 65 percent, but the biggest loser is South America. Vertebrate populations there have dropped by an average of 94 percent since 1970.[1]

The end point of population declines is extinction—when the last individual of the last population experiences what comes to us all eventually. Here too, the data don't look good.

Take the charismatic megafauna again, in this case birds and mammals. As we saw in the previous chapter, these are the species we know the best. Since 1500 CE, we've lost 159 species of bird and 85 species of mammal to global extinction. These include icons like the Great Auk, Dodo, and Thylacine (the dog-like marsupial also known as Tasmanian Wolf)—species we are never going to see again.[v] A further fifty species of mammal and bird are probably extinct—the Yangtze River Dolphin and Eskimo Curlew, for example. Seven more now exist only in captivity, like the Guam Kingfisher. Around one in six bird and mammal species are judged to be at risk of extinction within the next century, if things carry on as they are.

We can be confident that we know the extinction risks faced by the well-studied birds and mammals. For many groups, we don't even have names for most of the species. We know that we've lost twenty-three species of moth to extinction since 1500, but the true number gone is probably much higher. As for the number of species at risk of extinction in the near future . . . if the estimates of risk for mammals and birds are typical—and there are reasons to think that they are—and if there are about eight million species of animal and plant on our planet—again a

---

iv. These northern continents may be doing "better" since 1970 because they did worse before then, of course.

v. Because it's really hard to know when the last individual of a species has gone, we occasionally rediscover a species we thought was extinct. The New Zealand Storm-petrel is a recent example, relocated after having gone missing for 150 years or so. That's why the keepers of the list of extinctions (IUCN) wait fifty years after the last record before a formal declaration of extinction.

reasonable approximation—then it's likely that around a million species are at risk of extinction in the next century.

We know that extinction comes to all species. It's an end point as inevitable as your death. Do these numbers tell us any more than the obituary column tells us that people die?

Yes, they do.

Extinction is indeed a natural process, but what matters is the *rate* at which it occurs. Just as with the death of individuals in a population. Compared to the normal level of extinctions we see in the fossil record, current rates are of the order of a hundred to a thousand times higher. Probably nearer to (or above) the top of this range. For comparison, at the height of the Covid-19 pandemic, the probability of a person dying in the UK was essentially doubled. Now imagine what a thousand-fold increase would have felt like. Yet that is what we are looking at with the death of species. These numbers paint a very worrying picture.

A dramatic increase in death rates is a strong signal that something important has changed, and it's imperative to identify the cause. What's happened on our planet to cause the changes that we're seeing in the natural world—population declines, species extinctions, geographic distributions changing, alien species colonizing? I don't need to tell you the answer, do I?

I've been studying ecology for more than thirty years now. I was drawn to the field because I love the natural world and was motivated to understand how it works. Increasingly, though, the answers to my research questions come down to what people are doing. Humanity has inveigled its way into every mechanism underpinning natural systems. We are now a major contributor to every ecological process that determines the contents of my moth trap.

The basis of all ecology is the fundamental truth that all individuals are born and die, and that populations will only persist if deaths do not outnumber births.[vi] Widespread declines in natural populations around

---

vi. True of closed populations, technically.

the world—from vertebrates in South America to moths in Britain—are happening because humans are adding to deaths and subtracting from births.

Sometimes we add to deaths directly. Many species are in decline because of the extra mortality our actions have imposed.

We kill insects deliberately, and we do so very effectively. Estimates from 2014 were that around two million tonnes of pesticides were being put into the environment every year, of which around 600,000 tonnes were chemicals specifically designed to kill insects. Despite a lot of publicity about the negative impacts of all these pesticides on natural populations,[vii] use was expected to rise to 3.5 million tonnes by 2020. Obviously, much of this input is targeted against pests of agriculture, but chemicals do not always stay where they are intended. Many are sprayed onto crops as liquids, which then drift in the wind onto surrounding habitats. This drift deposits them in concentrations much lower than recommended for killing crop pests, but even low concentrations of insecticides—less than 1 percent of that recommended—can be fatal to moths and other insects. It's little surprise that natural habitats around crop fields have fewer moth caterpillars than comparable habitats farther away. Even so-called targeted insecticides like neonicotinoids, designed to be taken up into the tissues of crop plants to protect against insect herbivores, mainly end up elsewhere. As much as 95 percent of the application escapes into the wider environment, with widespread negative effects on bee and butterfly populations.[2] Remember that populations stop growing when the birth rate equals the death rate. Up the death rate, and that balance will occur at a lower population size—if balance happens at all. This means fewer moths to the trap.

Pesticides don't have to kill to reduce population sizes, though. Diamondback Moths are attractive dark brown micromoths, about the size of a grain of basmati, and patterned dorsally with a series of pale shapes of a form you can probably guess. Amazingly, these tiny creatures are migrants, regularly crossing the English Channel or North Sea to arrive in the UK in large numbers. Their caterpillars are also one of the most serious pests of cruciferous crops—cabbages

---

vii. In my echo chamber, at least.

and the like—and a key target of pesticides. Yet, while designed to kill, even sublethal pesticide concentrations have effects. They reduce the number of eggs laid by female Diamondbacks, with knock-on effects on the population growth rate. Reducing the birth rate can have the same effect as increasing the death rate.

Diamondback Moths number in the billions, and their enormous populations are concomitantly awash with genetic variation. This is the raw material of evolution. Some variants imbue Diamondback Moths with resistance to pesticides, and hence a huge survival advantage over their susceptible kin. Survivors beget survivors, and resistance rapidly spreads through the population.

Not all species are so lucky. Populations of Killer Whales in the seas around many European coasts now are small and declining. Their problem is a failure to reproduce. The small pod that lives in the waters of northwest Scotland and western Ireland has produced no calves in two decades of study. Those in the waters around Gibraltar do reproduce, but at a very low rate. These whales have accumulated very high concentrations of toxic chemicals called polychlorinated biphenyls (PCBs)—substances once commonly used as coolant fluids, and in flame retardants and waterproofing compounds. Production and use of PCBs have been banned because of their known toxic effects, but once in the environment they are slow to degrade, and these effects persist. The low birth rates in European Killer Whale populations are almost certainly due to poisoning by PCBs.[3] Without even trying, we've managed to create populations of potential mothers—in a long-lived and intelligent species—unable to raise young.

Sometimes we impose the conditions for higher death and lower birth rates, but the species adapts and bounces back. Other species do not have the resources to evolve. The future looks bleak for Europe's Killer Whales.

Humanity increases the death rate, and decreases the birth rate, of other species in many ways, but the most common way is through competition. We are extremely good at this.

Recall that ecologists think about different types of direct competition—exploitation, interference, and preemption. Humans outcompete other species through a process that combines two of these three: habitat conversion.

When humans move into a habitat, we will naturally consume some of the resources present. All animals consume, and we are no different. What we use will not be available for other species. As we have seen, when two species compete for a limited resource, the carrying capacity of the environment for at least one of the species will decline. That's if there's a clear winner—carrying capacity may be lower for both if they coexist. Assuming humans survive, we expect competitor populations to decline. Apples we eat are not available for Codling Moths. Classic exploitation competition.

These days, humanity rarely stops there. We've learnt that we can support much larger populations of our own species if we remove species that we can't consume and replace them with those that we can. In the first instance, this is exploitation competition on a grand scale.

We consume species, but we consume them with fire, or with machines. We burn whole tracts of forests or cut down their trees. We plough up grasslands, meadows, and heaths. We pump water from marshes and bogs, and then burn or plow over the vegetation left. The plants and animals we consume in this way are denied to the myriad populations that would otherwise have consumed them. The carrying capacity of the environment for almost all of these consumers—along with the species they consumed—has declined. Their populations can only decline, too.

The extent to which humanity has appropriated the environment is stunning. According to the United Nations, more than 19 million square miles is now used for agriculture.[4] That's just under 40 percent of all ice-free land worldwide. A third of this area is croplands, of which 10 percent, or 640,000 square miles, is permanent crops: fruit and nut trees, cocoa plantations, oil palm, and the like. The remaining two-thirds is meadows and pastures, largely for livestock to graze—though they also consume the produce of a third of the croplands. In effect, we've given over almost one-third of the available land surface of the planet to grow meat and milk. That proportion grows year on year. So too does the amount that we have built on, to house ourselves and provide transport links—currently 580,000 square miles.

Of course, we haven't completely pushed natural populations off this land, but we've dramatically reduced its carrying capacity for many species—entirely for some. Species like the original population of the Gypsy Moth in England, that vanished as its fenland refuges were successively drained and converted to farmland. Or the Cryptic Treehunter in Brazil. The forest habitat of this passerine bird has been extensively logged and converted to sugarcane and pasture. Tiny patches do remain, but evidently not enough to support the Treehunter. The species was only discovered by scientists in 2014, but it was declared extinct just four years later. Birds are relatively large and conspicuous. Who knows what more-unobtrusive animals and plants we lost for good in those forests? We will certainly lose more as the consumption continues. Orangutans are poster children for deforestation, but merely the tip of the tip of an iceberg of loss.

Having consumed natural habitats, humanity then shifts to wholesale preemption. Species are excluded from vast areas of land that they could use, thanks to our occupation of that space. Humanity has changed global vegetation patterns at unprecedented rates. We plant crops that most of our competitor species cannot use, and poison to oblivion any that dare to try. Herbicides, insecticides, fungicides: all keep the land sterilized of unwanted biodiversity. The richness of life that makes our planet unique in the universe is eroded and erased from vast and ever-expanding areas, and restricted to ever-smaller enclaves. In only a few hundred years we have moved from small pockets of humans surrounded by wildlife, to small pockets of wildlife surrounded by humans. If humanity retreated or disappeared, nature in some form could reclaim these spaces.

Competition for shared resources inevitably reduces the population sizes of any species that lose the contest, and even those that draw. We all know the perils faced by small populations. They can be knocked out by random episodes of bad luck, even if their birth rate would exceed their death rate in normal circumstances. Not only has humanity increased the frequency of small populations, but it has also increased the likelihood that bad luck will find them.

Think about that key component of environmental stochasticity, bad weather. The English now have greater recourse to this conversational gambit thanks to decades of pollution. Burning fossil fuels adds carbon dioxide (CO_2) gas to the atmosphere as a by-product. Plants absorb CO_2, and use the sun's energy to convert it to sugars, but we are now adding more than all the world's plants can remove. The concentration of this gas in air has reached more than 412 parts per million (ppm; 0.0412 percent). That doesn't sound like much, but the highest concentration in the previous 800,000 years was only around 300 ppm. This is a problem because CO_2 is a greenhouse gas—it absorbs and radiates solar energy, effectively trapping heat from the sun that would otherwise dissipate out into space. More heat means more energy in the atmosphere, with significant consequences for life on earth.

One consequence is obviously that temperatures are going up worldwide. We've already seen a global increase of over 1°C (1.8°F) on average temperatures relative to preindustrial times. But it's not the averages that matter so much when it comes to climate—it's the extremes. Heatwaves are happening more frequently, and heatwaves kill. They increase death rates in human populations, and they do the same for other species, too. The local extinction of plant and animal populations is strongly related to how much the maximum temperature they experience has increased over recent decades. We are already losing wild populations to climatic stresses that exceed the ability of their individuals to cope. Insects do better if there is natural habitat in which to take refuge from the heat, but of course we are rapidly destroying that.

More energy doesn't just raise temperatures. Energy drives storms, and more of it can increase their severity and frequency. Species with small and restricted populations may have no refuge in the face of severe weather systems. Take the Cozumel Thrasher, a species of bird found only on the small island of Cozumel off the east coast of the Yucatan peninsula in Mexico. Unfortunately, Cozumel often finds itself on storm paths, like Hurricane Gilbert in 1988—still Mexico's most severe hurricane. The Thrasher was quite common before Gilbert, but its population was obviously devastated by that storm. It quickly became rare. Further hits to the island from Hurricanes Roxanne in 1995, and Emily and Wilma in 2005, have brought the Thrasher to the brink.

There was one possible sighting in 2006, but none since. It might be hanging on somewhere on Cozumel, but the odds are against it. Again, who knows what more unobtrusive species we lost for good in those storms?

My perspective is very much biased to the terrestrial habitats favored by humans and moths, but Earth is really a water world. Aquatic species will suffer just as much as those on land from extremes of heat—as mass coral bleaching on the Great Barrier Reef and elsewhere proves.[viii] More problems arise from the basic physical truth that if you heat water, its volume will increase. It's easy enough to prove this with a saucepan filled to the brim. At the global scale, this expansion means that sea levels will rise and oceans will increasingly claim areas that we had taken for granted were ours. Coastal cities will suffer. So too will the animals and plants of shoreline habitats and low-lying islands.

Sea-level rise has already claimed its first species. The Bramble Cay Melomys was a small rodent found only on a tiny coral island in the Torres Strait, north of Australia. Although it was common in the 1970s, by the turn of the century its population had fallen to double figures, and the last individual was seen in 2009. Intensive searches since have failed to find it. Bramble Cay is a low-lying island, and storm surges sweeping across it killed both the animals themselves and the plants on which they relied for food. Bad luck for the Melomys. Such surges will only become more common worldwide as the oceans continue to absorb heat and expand, and as melting ice adds to their volume.

So not only has humanity pushed down carrying capacities for many species through competition for space and resources, but we've also upped the likelihood that environmental stochasticity will knock their remaining populations down to zero. You might have realized that humans were bad luck for wild animals and plants, but did you know how literally that's the case?

viii. Another problem is that CO_2 dissolves into sea water, acidifying it. This acidity dissolves coral skeletons.

Humans are consummate consumers. We have increased the death rates of many species by consuming their habitat. But we have increased the death rates of many others by consuming them. Predatory humans provide some of the most iconic stories of loss and extinction.

The Passenger Pigeon of North America was once the most abundant bird on Earth. Its flocks numbered in the millions, and it nested in colonies covering dozens of square miles. The species might have benefitted from safety in numbers when its enemies were falcons, snakes, or foxes, but the arrival of the shotgun turned this strength into a weakness. Hunters could kill several birds with a single round, and shooting Passenger Pigeons was so easy it was barely considered sport. Harvesting eggs and nestlings from its huge colonies was equally straightforward. It probably wasn't only hunting that drove this common species to extinction—habitat destruction may also have played a key role—but hunting certainly helped, and the last known wild bird was shot in March 1900. Its neighbor the American Bison (*Bison bison*) was lucky to escape the same fate, as hunting for subsistence and recreation reduced its numbers from more than 60 million to fewer than 600 in the course of the nineteenth century. Laws to save the last few were enacted just in time.

The most dramatic examples of people as predators come from oceanic islands, like New Zealand. This was the last major ice-free land mass to be colonized by humans, with the Maori only arriving around 700 years ago to find a land that was—a few species of bat aside—completely free of land mammals or snakes. As a result, the archipelago was a paradise for birds. Or at least it was until humans appeared.

New Zealand's birds had diversified into a variety of roles carried out elsewhere by mammals. Giant Moas grazed the forests, hunted by the largest birds of prey anywhere on the planet. Tiny wrens scuttled about the forest floor like mice.[ix] Many of the species had lost the power of flight after colonizing the islands, no longer having any need to take to the air to escape the terrestrial carnivores that had hunted their ancestors. They forgot what mammal predators were, and the

---

ix. The New Zealand Wrens are not closely related to the wrens of the Americas and Europe.

evolutionary memory of what to do when encountering one eventually disappeared too. This was bad news indeed when canoes bearing predatory mammals breasted the horizon.

Moa were so easy to catch that the Maori didn't bother to lug the whole bird home. The largest species probably weighed around 440 pounds in total, but the best bits were the yard-long drumsticks. Hunters just cut off the legs and left the rest of the carcass to rot. All ten Moa species had gone within about a century of first contact. The giant Haast's Eagle that depended on Moa for its diet went, too, outcompeted.

Unfortunately, humans weren't the only predatory mammals in those first canoes. They also shipped Kiore—the Pacific Rat. The sort of alien species that worries scientists around the world. While the Maori went to work on the large birds, the Kiore went to work on the small. Flightless or ground-nesting species were especially at risk from predation by these rats, and most of the New Zealand Wrens rapidly went extinct. Kiore are catholic consumers, though, and also impacted the native insects and plants. Remarkable groups like the Weta—large flightless crickets, some of which weigh as much as a House Sparrow—disappeared from much of the archipelago.

They weren't helped by the appearance of Europeans in New Zealand, following Abel Tasman's encounter with the islands in the seventeenth century. This second wave of colonists quickly decided that the islands would be improved by the addition of species from the Old Country, including predators like cats, stoats, and ferrets, and omnivores like pigs and hedgehogs. Black Rats, Brown Rats, and House Mice stowed away on the ships. These additional aliens ate their way through the medium-sized bird species that had been too large for the Kiore to cope with, but too small to be worth the Maoris' while. Species like the remarkable Huia, with its pronounced sex difference in beak shape—the short, stout chisel of the male and the long, curved forceps of the female.

Overall, New Zealand has lost more than sixty bird species to extinction in the seven centuries of human occupation, and most of these to the elevated death rates of predation. The islands are home to extensive tracts of native forest almost empty of native birds. A symphony of song forever lost. Other species have gone, too. The community of mammalian predators probably did for Buller's Moth, a huge New Zealand endemic with a 15-centimeter wingspan known from one individual collected in 1867. Even that specimen has since been lost.

New Zealand's forests are a living memorial to the power of predation as an ecological force. And the Law of Unintended Consequences. Humans didn't mean to cause all these extinctions. We just did.

It's not just New Zealand that has suffered to feed humans. The extinction of large, profitable prey species following the arrival of humans has been a story repeated the world over, be it Moa-nalo of Hawaii, Elephant Birds on Madagascar, Giant Wombats in Australia, or Giant Sloths in South America. Most of the world's megafauna has vanished over the last 40,000 years, coincident with the expansion of human populations. Our accompanying predators have wreaked equal havoc. It's estimated that domestic and feral cats kill more than a billion birds and six billion mammals every year in the US alone. Pet and feral cats have helped force the extinction of almost fifty species worldwide in recent centuries.

The numerous extinctions of birds and other species in New Zealand were largely unhappy accidents, but humanity actively promotes the extra mortality provided by predation to force down wildlife populations. One example is the Harlequin Ladybird, a brightly colored beetle introduced from East Asia to North America and Europe as a biocontrol agent for scale insects and aphids. Unfortunately, it's a voracious consumer of other ladybirds as well as its intended prey, and populations of many native species are in steep decline thanks to the Harlequin. Scientists are now looking at biocontrol agents for this biocontrol agent, aping the Old Woman Who Swallowed a Fly.

More examples come from attempts to limit that classic exploding population: Léopold Trouvelot's Gypsy Moths. As mentioned in chapter 1, early efforts to control this invader used Paris Green, a chemical toxic enough to kill rats as well as insects. The cons of spraying such a general poison widely across the environment were obvious, and attention soon shifted to insect-specific solutions. Entomopathogenic fungi and bacteria have both been used as more focused killers. We've already seen the power of parasitoids to drive down moth populations, and in 1906 one such insect—the tachinid fly *Compsilura concinnata*—was introduced from Europe to New England to control the Gypsy Moth.

Nowadays, rigorous checks are undertaken preceding the release of any biocontrol agent in order to ensure that it targets only the species we wish to suppress. A century ago, regulations were not so strict. It quickly became apparent that *C. concinnata* could parasitize species other than the Gypsy Moth. Indeed, it turns out it *had* to. Gypsy Moths have only a single generation a year, but the tachinid runs through three or four. The fly overwinters in a caterpillar host, but Gypsys overwinter as eggs. We now know that *C. concinnata* can parasitize a couple of hundred other species, including many North American native moths. It's not a great control agent for the Gypsy, but it does seem to have contributed to regional population declines in the impressively large Imperial Moth, a dead-leaf mimic and relative of the Silk Moth.[x] *C. concinnata* heavily parasitizes Imperial Moth caterpillars in mainland New England, but is absent from the offshore islands, that comprise the few locations where the moth continues to thrive.

Humanity is a highly catholic species, consuming members of many taxonomic groups. Yet we are still responsible for huge amounts of predation, even when we are not the predators. Those extra deaths drive population declines—often all the way to extinction.

Individuals are born, and individuals die. Between these bookends, organisms can choose to allocate what resources they acquire to growth and reproduction in a range of different ways. How evolution has shaped those life history choices contributes to the dizzying array of diversity in the moth trap, and beyond. The impact of humanity on births and deaths is driving change here, too.

These impacts are particularly strong when humans are predators. We tend to target larger animals—more meat per kill—and so exert an evolutionary selection pressure on prey to get smaller. This effect is most marked in species harvested commercially, such as Cod. In

---

x. Remember that Léopold Trouvelot's experiments with the Gypsy Moth came from his interest in silk production.

some Cod populations, body lengths at sexual maturity have halved, as fishing out of larger individuals means that investing more energy in growth risks missing out on the chance to reproduce at all. Fish in these populations are now starting to reproduce at much younger ages for the same reason. Smaller female Cod now devote more energy to ovarian tissue than they used to, and are much more fecund as a result. Larger Cod could—and did—lay more eggs, but there's no point in growing if death is going to strike before they get the chance.

The pressure of harvesting on larger individuals also impacts smaller species. The average sizes of several species of gastropod on the Californian coast, such as the Giant Owl Limpet, have decreased as a result of harvesting by the local human population. This could have significant impacts on limpet reproduction, as the species is a protandric hermaphrodite—smaller animals are more likely to be male, but change sex to become female when large. This gender-fluid strategy is advantageous because larger individuals can lay more eggs, but they still get to reproduce by producing smaller (and energetically cheaper) sperm while they grow. From the perspective of the individual limpet, it's win–win. If the larger females are selectively picked off by predatory humans, though, the advantage is lost. The population suffers twice—increased death rate and lower birth rate.

Mortality doesn't need to result directly from human predation to drive life history changes—other ways in which humans elevate the death rate are just as effective. The industrial scale of deaths from pesticides creates a strong evolutionary imperative for species to respond. Adapt or die. The result has been an arms race between humanity and the species we consider to be pests. We relentlessly seek new ways to kill as evolution neutralizes our weapons. Adaptation is not without its costs, though. Take the Oblique-banded Leaf-roller, a tortricid moth considered a pest of apples. One Canadian population evolved resistance to the pesticides used against it, but the resistant moths took longer to develop and ended up smaller. Size relates to how many eggs female Leaf-rollers can produce—as you will recall, it commonly does within and across moth species—so that the ability to survive pesticides comes at the expense of growth rate and fecundity. Humanity not only influences births and deaths, but how species live in between.

Humans are impacting natural populations worldwide, but populations do not live in isolation. Their distributions across the landscape layer to produce sets of coexisting species—ecological communities, of greater or lesser richness. We can debate the extent to which these communities are ordered ranks or random assemblages, but either way, the effects of humanity are being felt upon them. Because not all species are equally susceptible to our impacts, we are changing the composition of these communities.

Recall that communities develop through a combination of determinism and stochasticity. We've already seen how humans are changing the environment to increase the likelihood of bad luck, hitting populations in many cases already depressed through competition with, or depredation by, humanity. Determinism dictates which species will suffer the most. The British moth community is a perfect exemplar.

When the world is changing around you, it pays to have options. As a rule, specialists suffer. This is certainly true for British moths, where species with picky caterpillars are more likely to be in decline. It's why we tend to find fewer specialists in more-degraded environments, like the London Borough of Camden. Specialists also tend to have smaller geographic distributions than generalists, and so are already starting from a position of disadvantage. Larger, slow-growing species will find it tough as well. Better to develop quickly and exploit resources while you can. Short flight periods and the inability to relocate to potentially better sites likewise leave species vulnerable to uncertain and changing environments. All of these traits are common in species being lost from moth communities across Britain.

When the world is changing around you, it also pays to have backed the right horse. In these days of rapidly warming climates, we see winners and losers in British moth communities broadly falling out in terms of their temperature preferences. Some like it cold, and these species are generally in decline. For example, the Red Carpet, a brightly banded moth of higher altitudes and latitudes in the UK, is decreasing particularly sharply at the southern margins of its range. Their loss is warm-loving species' gain. As the climate ameliorates, more-northerly areas are becoming hospitable to species that could not survive the cold weather previously typical of those regions. Humanity creates winners as well as losers.

Moths that reach the northerly limits of their distributions in Britain are generally spreading north, at a rate of around 5 kilometers (3.1 miles) per year. The largely southern Double-striped Pug, one of the commonest moths in and around my trap in London, is now expanding its distribution through northern England and over the border into Scotland. Its range grew 165 percent in the period 1970–2016. We've already met continental colonists like the Tree-lichen Beauty and Vine's Rustic that have taken the opportunity offered by a warming climate to cross the English Channel. The overall increase in the number of species in the British moth community is due in part to the fact that warmer climes are home to richer assemblages, and hence more species are available to colonize as our climate warms. Extended warmth across the year also allows species like the Double-striped Pug to pack in more generations. Historically double-brooded in southern England, its first brood is now emerging earlier, allowing a third brood to be squeezed in before the end of autumn. This Pug more than doubled in abundance between 1970 and 2016.

However, species that can tolerate a warming climate may still suffer if their food suffers, or their relationship to that food.

Moorland moths like the tweedy Gray Mountain Carpet are a group of species undergoing some of the most significant declines in the UK. Climate is one likely cause, but other changes to their habitat are playing a part. More-frequent burns—a way of promoting young shoot growth for the benefit of grouse, and grouse shooting—and heavier grazing by livestock may be degrading the habitat. Artificial fertilizers are also having an impact. Even when not directly applied to a habitat, their overuse in agricultural areas produces atmospheric nitrous oxide as a by-product. This pollutant can then rain down onto, and fertilize, nutrient-poor soils, to the detriment of specialist plants of such habitats, like the Heather and Bilberry on which the Gray Mountain Carpet caterpillars depend. Such pollution is one way in which habitats can be degraded, and communities changed, even when apparently well protected. All of nature is connected.

The Common Quaker is one of those species that could be mistaken for the Uncertain, and this also describes its future. Quaker caterpillars feed on the young leaves of oaks and other trees. Climate change means that trees are coming into leaf earlier, before the caterpillars are hatched and ready to feed. Quaker caterpillars grow more slowly on less palatable older leaves, giving predators more time to find them, and eventually grow into moths more likely to show developmental abnormalities—vestigial wings that prevent them from dispersing and finding mates. The Quaker population declined by almost 20 percent in the period 1970–2016, and this growing mismatch with its food might be driving that trend. The Winter Moths we met in chapter 2 are facing similar problems. The knock-on consequences for species that depend on these moths for food, like the Great Tit, are worrying, too.

Communities have always been protean. Of course we expect them to change. We just don't expect a single species to be responsible for most of the change happening across the whole planet.

The composition and structure of ecological communities doesn't only depend on what happens in their immediate vicinity. Events in the wider environment are important, too. All of nature is connected. This is why migration matters.

Indeed, migrants have never mattered so much. Humanity has destroyed a substantial proportion of natural habitat worldwide, and much of what is left is now heavily fragmented—small islands in a sea of inhospitable cropland, pasture, or concrete. The populations they house will be small, too, and susceptible to the vagaries of bad luck. Luckily, as we've seen, fragmented populations can still persist if they are connected by migrants. Migrants can bolster birth rates and counteract death rates, preventing population extinction and recolonizing sites when local extinction does take populations out. Humanity's fragmentation of nature has only increased the relevance of these dynamics.

Migration can ameliorate some of the damage caused by fragmentation, but only some. Metapopulations are most secure when there is a large "mainland" population acting as a plentiful source of immigrants.

Unfortunately, habitat destruction tends to reduce the extent and productivity of such mainlands, to the detriment of surrounding patches dependent on their largesse. Remaining fragments are often viewed as unimportant from a biodiversity perspective, but destroying them can increase the distance between surviving patches, and so lower the likelihood of colonization. When colonization rates are lower than extinction rates, populations will eventually disappear. More isolated habitat fragments have fewer species, moths and others.

On top of that, not all species are well adapted for a peripatetic lifestyle. Female Vaporer Moths, for example, lack wings, essentially being furry sacks for laying eggs. They are ill equipped for moving between habitat fragments. Likewise, Winter Moth, Mottled Umber, and Early Moth—all widespread species I've trapped in Devon but not in London, where the patchy nature of suitable habitat does them no favors. Even apparently mobile species often will not move far, like the Cinnabar Moth we met in chapter 2. Many skulking bird species of the Amazon rainforest understory evidently will not cross open spaces to the extent that major rivers in this basin become boundaries to their geographic distributions.

Specialists on certain habitats or food plants will fare especially badly when fragmentation increases. Species like the Scarce Pug, which in Britain feeds only on Sea Wormwood on a few east coast salt marshes. Extensive coastal development means that salt marshes are rarer and more-fragmented habitats than of old, and these are the only habitat of Sea Wormwood in Britain. Greater distances between suitable patches reduces the chances that dispersing individuals will find them, to colonize or rescue.

Migrants can also allow species to respond to changes in conditions— to take advantage of new opportunities as they develop, or escape from sinking ships. This is especially important in the face of the ongoing climate crisis. When environmental conditions change beyond the physiological tolerances of individuals, the species has only three options: adapt, move, or go extinct. The current speed of environmental change makes adaptation difficult, especially for those with slower life histories, leaving movement as the best option for survival.

Unfortunately, the ability of species to track changes in the climate is significantly hampered by habitat destruction and fragmentation. It's easy for populations to move through continuous tracts of habitat. But remember the effects of area and isolation on the species richness of islands: small, remote pockets of habitat are harder targets for dispersing individuals to hit. Humanity has increased the need for species to move while simultaneously making it harder for them to do so.

We can help, though—right? If species need to move, we can step in and do the leg work. It's called *assisted colonization*—the translocation of individuals beyond the current limits of their distribution in order to conserve species that would otherwise go extinct thanks to their inability to reach new areas in the face of a changing environment. Humans have been moving species around for all sorts of reasons for millennia now.[xi] Why not for conservation?

Well, precisely *because* of those species we've moved—the impacts of pesky aliens like the Box-tree Moth. In truth, that species is second division when it comes to damage. Other aliens have been much worse. I've already mentioned cats and rats, but take the Rosy Wolfsnail. It was moved to several islands across the Pacific to control populations of another alien, the Giant African Land Snail, but instead ate its way through the entire world populations of more than 130 other snail species. Alien diseases can wipe out naïve host populations, like the fungal pathogens *Batrachochytrium dendrobatidis* and *B. salamadrovirans* that, between them, have been responsible for the extinction of almost 100 amphibian species worldwide, and population declines in hundreds more. Alien plants can modify ecosystems to their own advantage, and suppress native plant species. Native birds tend to do worse in habitats dominated by alien plants, because their insect prey often cannot make a living on those plants. Aliens in general have been associated with the global extinction of more species in the last 500 years than any other human intervention, including habitat destruction. They remain one of the main drivers of global population declines.[xii]

It's trebly ironic that not only has humanity caused problems for species by increasing the need for them to move while simultaneously

---

xi. The earliest known translocation was a marsupial, the Gray Cuscus, from New Britain to New Ireland around 20,000 years ago.

xii. Like that experienced by Blackburn's Sphinx Moth. First collected by Rev. T. Blackburn in the 1870s, my namesake is now on the US endangered species list. The main ongoing threat seems to be predation by a variety of alien insects.

making it harder for them to do so, but also has caused problems for some species by moving others. The pressure for assisted colonization is growing, but we are rightly wary of taking species to places where they have no prior history.

The contents of a moth trap are the culmination of a drama that has been playing out across the four billion or so years that life has existed on this planet. The environment is the backdrop and ecology writes the script. The cast, on the other hand, is ultimately chosen by evolution. Even here, humanity is having its say.

The number of species with which we share our homes, countries, regions, and planet is ultimately the outcome of speciation minus extinction, with immigration also thrown into the mix. It should be obvious by now that we're seriously affecting the extinction rate. It's 100 to 1,000 times above the rate we'd expect, except during mass extinction events like the one that claimed the dinosaurs. It's sobering to think that we're as good for the planet as an asteroid strike. All those aliens tell us that we've boosted the immigration rate for most biogeographic regions to a similar degree. In fact, the UK now has as many alien plant species as native. Hawaii has more established alien than surviving native bird species. The last time we had such a substantial biotic exchange was when North and South America touched fingers in the Isthmus of Panama. That only concerned those two continents. Humanity has boosted immigration rates worldwide.[xiii] Around the world, most species are not alien—but this is still a massive shift.

Whether humans have altered speciation rates—and in which direction—is much harder to say. Most speciation probably happens in allopatry, when populations diverge after becoming separated by barriers to dispersal. Humanity has split many populations through our destruction and fragmentation of habitats—like that of the small primate Geoffroy's Tamarin divided in two by construction of the

---

xiii. Those gains don't compensate for the losses—extinct species and aliens are generally very different in form and function, like replacing the Prince of Denmark with another Yorick.

Panama Canal. We're already seeing genetic differentiation in the Tamarins on opposite banks. On the other hand, as we reduce the populations of most species, we also reduce the quantity and diversity of genetic material available for evolution to work with, the very material that underpins speciation. Whether the fragmentation and reduction in populations yields increases or decreases in speciation rates is an open question.

We certainly have been responsible for new species arising, though. Our penchant for moving organisms around allows them to encounter and exploit new opportunities, and potentially diverge from their original stock. Those aliens can themselves provide opportunities for other species to diversify. Recall the European Corn Borer, which has only bored corn since we put that crop in front of it, but which is in the process of splitting into two species as a result. New species can also arise when alien and native get together. A classic is the Common Cordgrass. This plant is the hybrid child of European native Small Cordgrass and North America alien Smooth Cordgrass parents, which first met in southern England in the nineteenth century.

Humanity is not only playing with the cast, but also changing the stage upon which evolution shapes those actors. We know that how speciation, extinction, and immigration add up is shaped by environmental conditions—conditions that we are altering. Our impacts are affecting the very processes underpinning biodiversity.

Global variation in species richness is determined by a combination of time, area, energy, and stability. More of all of these has tended to produce regions with more species. While there is little that humanity can do about the march of time,[xiv] we are changing the other three drivers of diversity, and with worrying consequences. Our appropriation of habitat effectively reduces the area available for species to persist and diversify. The colossal scale of our pollution of the atmosphere is adding energy and instability to global systems. While more energy has given us more species in the past, the speed at which we are adding it will certainly be bad news for many species in the immediate future. Not least for us.

---

xiv. Though we can still accelerate the rate at which processes run through it . . .

Our planet has naturally cycled through warmer periods, but what is different now is the rate at which we are transitioning to a Greenhouse Earth. Species can only move, adapt, or die. Not only have we reduced opportunities to move, thanks to habitat destruction, we have also reduced the time available to adapt. Humanity's current behavior looks set to increase global average temperature by around 3°C (5.4°F) by the year 2100. If so, a high proportion of species will soon be exposed to conditions beyond their current physiological tolerances. Moreover, that exposure will hit many coexisting species more or less simultaneously, because species adapted to certain environments will have similar tolerances. Some species may be able to adapt—those able to cycle through lots of generations in the time available. Many will not. The result will be an abrupt collapse of ecological communities. It will be worse in the tropics, home to most diversity, because tropical species are already closer to their upper thermal limits. If we do not rein in our current behaviors, in terms of greenhouse gas emissions and environmental destruction, we may start seeing these collapses in the next decade.

We don't know what will replace the ecological communities that collapse. What we do know is that we are part of these communities, and we depend upon them. Will we move, adapt, or die? For many of us, the choice will be limited.

That first moth trap in London got me thinking about what we know about the workings of the natural world. To understand the contents of this small box, we need a global view: they depend on the workings of all of nature. Now, it's hard to open a moth trap and not to see the fingerprints of humanity all over those workings.

Some are obvious. That first Box-tree Moth was a smack in the face for an ecologist, especially one interested in biological invasions. It should have been nowhere near a moth trap in urban London. It was an immediate reminder: there is no such thing as untouched nature. Likewise, the Horse Chestnut Leaf-miner that same morning—an alien moth born from an alien tree. And the Gypsy, from a reintroduced population of a species driven extinct in the UK by humans. Our fingerprints all over the trap.

Others were a slower burn. I was excited about catching thirteen Tree-lichen Beauties until it twigged that this was no longer a rare immigrant. Now it was one of a growing number of colonizers heading north in the wake of a warming climate. Ditto the dozen Jersey Tigers, a species I'd thought confined in England to south Devon. These were the two most abundant species in that first trap, both likely there because humans are changing the environment they depend upon.

Taking the trap to the Devon countryside just emphasized our impacts. Yes, mothing was better there. More individuals of more species, a richer community. But that was simply to highlight the relative paucity of the London trap. I'm excited to catch as much in London as I do, but the bar is low. And identifying species in the Devon trap led me to their backstories. Many species are in widespread decline, like the Orange Moth and Beaded Chestnut. Their appearance in the trap is perhaps a temporary pleasure, before the downward slide of their populations takes them away from the UK entirely. The multilayered and synergistic impacts of habitat destruction, fragmentation, pesticide application, urbanization, and climate change are setting many of the moth species I catch (and don't) on the path to extinction, locally if not yet nationally. Our fingerprints, again.

My moth trap is a source of joy to me, but it's ultimately a tool for sampling the environment. By piecing together the snapshots it takes of the hidden world of moths, we reveal what that world is like. The image we get is not a photograph, though, but a movie. As the frames flick past, we see the image changing. Even the longest-lived among us only get to experience a tiny fraction of the whole production, but we still see the actors come and go, and the narrative develop. We don't have to watch for long to know that for many of those actors, it's not going to be a happy ending.

The natural world is endlessly fascinating and beautiful, but humanity is grinding it down. We know this. Yes, not everything will go. Our interventions in the processes that drive the ebb and flow of populations, communities, and species will create winners as well as losers. But the biggest loser? Spoiler alert—it's us.

Chapter 9

The Acer Sober
Conclusion

*. . . It is also being forced upon our reluctant attention that the species
Homo sapiens is no privileged exception to the general conditions
that determine the destinies of other living species.*
— H. G. Wells

Acer Sober, London.

A moth trap is a constant source of wonder and delight. I have been captivated since that first morning at the Kindrogan Field Studies Council estate, when I discovered the array of flamboyant colors and extraordinary forms that could be magicked out of the darkness. The revelation that the same trick could be pulled on a roof terrace in urban London turned captivation to obsession. I'm more likely now to check the weather forecast for night than day. Any holiday or trip is predicated on opportunities to run a trap. Place and time are always triangulated with the question of what might be flying that night. Mornings now are greeted with an immediate frisson of excitement—what jewels will I find when I open the box? Hundreds of other people are sharing that experience up and down the country, and at different times, in other countries around the world. Hoping for something mega—rare or large or gaudy, or all three—but just happy with whatever the night has brought. To start the day surrounded by a sample of life's delicate beauty.

Running a moth trap is illuminating. In many ways.

Mothing is a hobby for me, but I have the extra privilege to be an ecologist by profession. It's my job to help humanity understand why Earth is home to such a dazzling array of biodiversity, and to introduce more people to its wonder. The moth trap is a means to both of these ends, and one that has done the same job on me. Why does it catch the species it does, in the numbers it does? How do those numbers vary, and why? I'd been thinking about questions like these on and off for years, but the moth trap refreshed my interest in the answers.

My life has been illuminated by trapping moths, but so too has my appreciation of their lives.

Ecologists think of ecology in terms of individuals, populations, communities, metacommunities, and the interactions within and between these levels of organization that determine variation. All of that is underpinned by the driver that is evolution, the process without which nothing in biology makes sense. My questions about moths need answers that reach from a morning on my roof terrace to embrace the complete geography and history of life as we know it.

Four billion years ago, give or take, there appeared in the ocean an entity of a different type to anything found on our planet before. Self-contained within a membrane, it was capable of powering itself through metabolism, and replicating itself to produce more of the same. It was the first entity we would classify as alive.[i] That first organism would have replicated simply by splitting into two, but we know the potential of reproduction. It made new copies of itself faster than those copies were destroyed, and its numbers increased.

Reproduction is life's superpower, but no population can increase forever. All organisms require resources for power, growth, and reproduction, and on a finite planet—or in a limited part of it—resources will run out. The birth rate slows, the death rate grows, or both. Inevitably, untrammeled expansion turns to a struggle for existence. Kin compete with kin for the scraps.

Numbers stop increasing. If the population has overextended itself, or destroyed its resource base, numbers start to fall. From these fundamental truths flows all of ecology.

Replication is not a perfect process, and not all individuals in that first population were identical. Variation like this is the raw material of evolution. Some individuals did better than others in replicating in times of feast or times of famine, and those were favored in the numbers game. New variants opened up new opportunities in the environment, leading to new populations. The process of divergence had begun—subdivisions promoting the development of new forms, leading eventually to what we call different species. Ecology and evolution interacting, to set life on the inexorable path to diversity.

The four-billion-year story of life has had uncountable twists of plot. Some lines in the narrative led to the colonization of land. One allowed the arthropod ancestors of insects to leave the oceans. Another

---

i. It's difficult to be sure about anything this far back in time, and this glosses over a lot of uncertainty. These are the key elements of life, though. I've assumed that life only arose once, but it's also possible that it appeared and disappeared some number of times before finally taking hold.

landed the vertebrate ancestors of mammals. Three hundred million years ago, a population split to give us one sister that was the ancestor of caddis flies, and another sister that was the ancestor of moths. About halfway through the radiation of moths, a population of mammals split into sisters, the ancestors of marsupials and placentals. The twists have continued, life after life, split after split, giving us a cast of characters we cannot even enumerate. Uncountable organisms, divided up among unknown numbers of species. All connecting back to that first entity, through an unbroken chain of births. One in nine of the species a moth (or butterfly), and one unusual species of mammal able to discern process from the mind-boggling abundance of pattern.

The world has not been created equal when it comes to this diversity. Speciation plus immigration minus extinction sums to the richness of each biogeographic region. The greater the dimensions of time and space, the greater the availability (up to a point) and reliability of energy, the higher the answer to this sum, and the more species we find in a region. The rules of diversification over millions of years on an ever-changing planet govern our expectations for the outcome of a night with a moth trap. We might expect hundreds of species in the tropics, where diversity is high, and dozens in Europe or North America, where species numbers are more modest. This is the context into which we shine our light.

Regions have not been created equal, but inequality is also the order of the night within them. Some spots are just better than others when it comes to species numbers. Order and chance both play a role: a greater diversity of resources allows a greater diversity of users. No species can be equally competitive in all circumstances, and so each has its preferred requirements. Quality of the environment matters for each species, but so too does quantity. More habitat is better than less, because smaller patches will support smaller populations. A bit of bad luck can knock them out. Specialists are especially susceptible, as they usually have fewer options. Quality and quantity are why my trap comes alive when I take it to Devon, where there are more individuals, and more species.

Diversification determines the set of species from which the community is constructed. Migration links the two. More and better habitat makes for richer ecological communities, but so too do better connections. Chance can subtract species, but migrants can add them back in. Much of the planet would be bare rock if it weren't for immigration, and isolated habitat patches would be less diverse. Leaving home in search of new opportunities is inherently risky, but larger and closer targets are easier to hit. Migrants can compensate for shortcomings in the quality and quantity of habitat, but quality and quantity make their task easier. It is movement that brings moths to my trap, even in the midst of one of world's largest cities. Some come from neighboring gardens, others from neighboring countries. Ultimately, all of nature is connected.

Regions and communities are not created equal, but inequality extends further still—to the species themselves. There is no one right way to be a moth. Most are small and ephemeral creatures, the adults spending a few days or weeks on the wing, completing their primary function—reproduction—as soon as they can. Most of the eggs they produce don't make it to adulthood, but species look for ways to improve their odds. For most, that involves hurrying through development as fast as possible, to minimize the time that death has to find them. Some find ways to protect themselves from death's sickle, like the Goat Moth that passes long years buried deep in the solid and reliable refuge of heartwood. Others find protection in chemical weapons, advertising their defenses with bright colors or cautionary calls. Some don't waste energy on fripperies like wings when it could be invested in eggs, and their offspring inherit the family tree. Others cross continents in search of opportunity. Moths have spent 300 million years finding diverse paths to negotiate their allotted time. The moth trap catches the glory of this journey.

Life is brutish and short for most because no species is an island entire of itself. All animals are consumers, but most are also consumed. A huge variety of creatures live on moths, from viruses and bacteria, through spiders, wasps, and beetles, to birds, bats, and even humans.[ii] Most moths never make it to the wing. Even good times tend not to last, because any population that flies too high will eventually get

---

ii. That famous bush tucker dish Witchetty Grub is the caterpillar of *Endoxyla leucomochla*, an Australian relative of the Goat Moth.

its wings clipped by predators. They keep populations in check, drive fluctuations in what the moth trap catches, and promote coexistence. Billions of moths are consumed every day, and while I lament those that the Robins and Great Tits steal from around my trap, diversity supports diversity. Predators are in part why we see the land as green. To be an animal is to consume, and much of nature's beauty is built on the foundation of slaughter. A moth trap's catch is witness to both.

All animals are consumers, and so what they consume is likely to matter as much as what consumes them. Populations can be steered from the bottom up as well as the top down. The security of a consumer population depends ultimately on the security of its food, and the dietary preferences of the moths in my trap tell me much about the local botany. I'd know there was a Horse Chestnut near my London flat, I'd know Devon was good for lichens, even if I didn't know. The moths tell me. But the presence of food doesn't guarantee the presence of a consumer—when more than one species is in the market for a meal, competition will result. Then it comes down to win, lose, or draw. Species will only coexist if they can find a situation where the greater struggle is with their own kind. If their competitor is always better, they will lose. The moth trap shines a light on which have succeeded—at least in its immediate vicinity.

The life of a moth is the product of rules. Rates of birth and death, and the roles of competition and predation; the division of resources among growth, survival, and reproduction; stabilization and equalization, migrants as colonists and rescuers; the effects of time, space, and energy on diversification.[iii] They work together to drive the richness and coexistence of species, to determine where they live and in what numbers. The basic definition of ecology. The core of the answers to the questions that the moth trap poses.

There is beauty in these rules, but life is not only shaped by rules. For me, it's the role of chance that really strikes awe.

For all the power of reproduction, the redemptive gifts of migration, and the ingenuity of natural selection, every creature we see around us only exists because its ancestors won the lottery again and again and

---

iii. And others I've ignored for both our sakes.

again. They survived to reproduce where most failed. They struggled through asteroid impacts, catastrophic global cooling and heating, and oceanic acidification and oxygen depletion that eradicated millions of species. They weathered extremes of heat and cold, dodged storms and droughts, evaded predators, survived epidemics, located new resources, bounced back when their numbers were small. They were blown out of open windows. They fluttered up into the warmth of a London night and were dazzled by the light from a fluorescent bulb. Rules define the patterns of life, but it is luck that has colored them in. Gypsys, Footmen, Eggars, Goats Uncertains, Silver Ys, Hawks, Box-trees, and Darts: time and chance happened to them all.

Every morning I wish for luck as I approach my moth trap. I'm lucky to be there to do it.

A moth trap shines its light on only a small part of the world, but the animals it draws in illuminate connections that span the entire globe.

Without those connections, Water Veneers would not survive long in my local pond. Vine's Rustic would not have colonized London on the slopes of a warming climate. Silver Ys would not grace summer with their gentle hum. We cannot fence off our own little corner of the world and expect it to thrive, or even to survive. However we protect it, it will fall into poverty. All of nature is connected, and only by ensuring those connections will it prosper.

The obverse is that we cannot insulate our small corner of the world against what is happening elsewhere. We appreciate that the actions of our neighbors affect us, but appreciate less that we are all neighbors. Losses over the horizon matter just as much as those in our immediate vision. Wherever the natural world is destroyed, all of nature is ultimately the loser. Every loss knocks another hole in the hull of the ship on which we sail. Connectedness shores us up, but can also bring us down.

The most illuminating thing about running a moth trap, though? It's the attention it draws to how tenuous are the threads by which nature hangs.

Sometimes a trap is heaving with moths, and other times it is empty.

We know the rules that affect these numbers, and we know the role that chance plays. But we also know that our actions are snipping away at the threads. The fate of much of nature is in our hands. That fate will be our fate, too.

Homo sapiens is an animal like any other. It's a consumer like any other. Do you think we are exempt from the rules by which the rest of nature plays? Think again.

You are born and then you die, and in between you will divide what resources you can accrue among growth, survival, and reproduction. We have already seen this division changing as the likelihood of early death has decreased worldwide. Resources are finite, though, and the larger the population grows, the fewer there will be to go around. We know what happens when a closed population reaches the limit of its resources: the birth rate drops, the death rate increases. When a population outgrows its resources, the only way for that population is down.

Humanity has been propping up the growth curve, thanks to our unusual ingenuity in providing for ourselves[iv] and fending off the attentions of those that would consume us,[v] but is only postponing the inevitable. We are already using Earth's resources unsustainably. Instead of living off the interest of our planet's natural capital, we are eating into the capital itself. Every year, the overspend increases. In our use of fossil fuels, we are burning the past to support our burning of the future. The tiny island in space on which we sit is rapidly being whittled away.

We seem to think we can live without the services nature provides. We can't.

Plants oxygenate the air and remove carbon dioxide. Forests absorb solar energy and transfer water from soil to atmosphere, affecting temperature and rainfall, and modifying the weather. Plants turn sunlight into leaves, stems, flowers, fruits, nuts, and tubers that we can harvest to eat, and fibers and wood for clothing and fabrication. They feed the animals we eat, milk, wear, and pet. We use the chemicals plants produce to cure headaches and cancer—and who knows what cures or other benefits lie hidden in the tissues of the vast majority

---

iv. Scientists can claim some of the credit for this.

v. The global pandemic of 2020 and 2021 notwithstanding.

of species we have not yet studied? Moths, flies, bees, birds, bats, and others pollinate plants—without them, many plants would not be able to reproduce, and we would be starved of their fruits and nuts. Vertebrates and invertebrates disperse seeds, allowing new plants to grow. They consume herbivores that would otherwise multiply out of control, stripping plants (and our crops) of their foliage. Dying plants add organic matter to the soil to fertilize the next generation—but this matter would not decompose without bacteria, fungi, or invertebrates to break it down and aerate the soil. We would have no soil for our plants without them. Yet other bacteria, fungi, invertebrates, and scavengers dispose of the feces and bodies of animals, reducing the risk they will fester and spread disease. Nature adds beauty and meaning to our lives, the inspiration for art, literature, music, science, and technology. We can't live without it, but more than that—it makes our lives worth living.

The current state of nature is dire. The numbers of moths coming to British traps has been falling for decades. Moths are far from alone in their declines—the majority of wildlife populations around the world are heading inexorably down. We are destroying nature remorselessly. It's impossible to run a moth trap and not be confronted by this.

The moth trap truly is a light in the dark. A warning light. It's a light we need to wake up to.

The first role of a scientist is not to have answers, but to have questions. The moth trap has posed many questions to me from my first tentative investigations of its contents, and I have posed some of them to you. I hope I've also given you some insights into how the answers might look—a scientist's second role. The implications of some of those answers are undeniably depressing. So I have one final question, one I'd like you to ask yourself.

What can *I* do to help the natural world?

This is about the most urgent question we have right now, given the many attacks that humans are raining down upon our own life support system. Fortunately, there is no shortage of answers. If we act, if we all

make changes, we can transform our fate. Because make no mistake—
the fate of nature is our fate, too.

First and foremost, we need to recognize ourselves for what we are—
animals. We are not the same as other species—no species is—but we
share much in common with them. First and foremost, our world runs
on energy, just like that of every other species. Life is hard work, and all
work needs energy. To get our energy, we consume. We *have* to—there
is no way around it. That makes us competitors, as our consumption
denies space and resources to others. Ecology tells us the consequences
of consumption and competition. If you want to help nature, make
choices that reduce the amount of land you need to support yourself and
others.[vi] If you want to help nature, first and foremost, use less energy.
Consume less.

The best way to be less of a competitor is to be less of a predator.
As we've already seen, humanity has given over almost a third of the
available land surface of the planet to grow meat and milk. We grow
crops to feed to livestock that we then eat, when it would be far more
efficient to use that land to grow crops for us to eat ourselves—habitats
can support a far greater biomass of herbivores than carnivores. Much
of the land we use to grow meat is in the highly diverse tropics. Every
hectare we use there puts us in competition with more species than an
equivalent area at higher latitudes. Not all farmed land is suitable for
growing crops, but that doesn't mean we have to use it for livestock.
Some of that is land we can return to nature. Anyway, many people
eat more protein than they need, particularly in the richer nations. I'm
not an evangelist for veganism, but the more plant-based your diet, the
less your impact on our finite land area will be.[vii] Cut down on animal
protein if you can't give it up. You'll save money—and I suspect you will
miss it less than you expected.[viii]

---

vi. There are lots of practical suggestions for how to do this in *Rebugging the Planet*
by Vicki Hird.

vii. The same arguments apply to animal protein derived from marine and freshwater environments.

viii. I appreciate that this is written very much from the perspective of citizens of
rich nations. But then it's those citizens that are *by far* the largest per capita consum-

Humanity appropriates land not just to feed ourselves, but also to extract materials for clothing, furnishings, possessions, transport, and power. There are so many opportunities to consume less here. Cotton alone accounts for around 35 million hectares (86 million acres) of the land we exploit, again mainly in the species-rich lower latitudes. Every T-shirt or pair of jeans you don't buy means less competition with nature.

As well as consuming less, we can reduce our impact through what we do consume. Not all agriculture is equally damaging to wildlife, and so it helps to support that which treads more lightly. In general, this is food that is grown locally, with less pesticide and artificial fertilizer, in more-diverse cropping rotation systems, in smaller fields, with more natural habitat interspersed. We do not have to erase nature from the land and grow huge monocultures of the food we need. It's better to share. By introducing diversity into our crops, we reduce the likelihood that large pest populations will develop. By promoting natural predators, we co-opt wildlife to control those that do develop. By encouraging pollinators, we increase yields of those key crops they fertilize. By promoting diversity, we increase functional redundancy. Generalist species are usually more robust to environmental stochasticity than specialists—by introducing diversity to our diets, we inure ourselves by becoming more generalist, too.

We can apply the same principles to those spaces we have direct control over—our gardens, roof terraces, or window boxes. It doesn't take much to share our space with nature better. We can start by imposing less mortality from pesticides, herbicides, and fungicides. Leaving lawns to grow wild in summer provides food and shelter for moths and other animals. Fallen leaves similarly—don't rush to rake them up. Wildlife benefits from habitat diversity, so provide it. Ponds, log piles, compost heaps, bee hotels, trees—all will add variety to the wildlife in your garden. Let your window boxes run wild, too. Grow native species before aliens. Cut out light pollution—artificial light at night

---

ers of the planet's resources, and for whom the need to consume less is most urgent. As a general rule, the better off you are, the more you need to cut your consumption.

has substantial negative impacts on the abundance and development of moth caterpillars. All of these measures ultimately save you work, and money, too. And there is so much more beauty in a riotous tangle of insect-buzzed blooms than in a silent, monotonous buzz-cut lawn.

As well as providing the building blocks of diversity, your efforts add stepping stones to facilitate its spread. You are partaking in metapopulation dynamics—providing shelter and fuel for migrants as they colonize new habitat patches, or rescue populations already present. It's the very essence of thinking globally by acting locally. Even small islands in an inhospitable sea can make a difference to the survival of individuals, and the linkages they provide. Grow those islands.

Competing less for resources and space has positive impacts on climate change. Artificial fertilizers and their production are major sources of greenhouse gases. So too is the production, packaging, refrigeration, and transport of animal products, in particular those imported from overseas. Whatever you eat, eat local produce whenever possible. Habitat destruction to support our consumption liberates carbon into the atmosphere, whereas habitat restoration removes it. Consume less to compete less—and give nature a better chance of helping us solve that other existential threat to our civilization. Humans are the same as other species, but also we are not. The key difference is our ability to recognize the consequences of our actions before they run through to their inevitable conclusions.

A fungus protects the algae that feed it, and lichens festoon the trees around my Devon moth trap. Moths consume leaves and nectar, but also pollinate, helping to propagate the plants on which they depend. We need to recognize that we are moths. We can no more survive without a healthy environment than can they. We need to recognize that the cycle of consumption and destruction, and all the other iniquities inherent to the current human ecosystem, are not inevitable. We need to work together to overcome those who would have us believe that they are.

We are animals, and as such we must consume, but many of us have come to let that fact dominate our lives. We need to consume less, but also to become more than consumers. All creatures make decisions about how to allocate precious resources over the course of their life history. If you're not working only to consume, what might you do with yours?

The middle of July 2020, and after having been locked down in rural Devon for four months, we had been back in our London flat for a week. It was almost exactly two years since I had first run a moth trap on my roof terrace, and I was finding the return to that aerie depressing. Our time in Devon had coincided with the sunniest English spring of my lifetime, and one of the warmest. Trapping moths there had spoiled me. Through May and June, I was regularly picking more than 300 individuals of forty-plus species out of a heaving trap. Against a backdrop of a world in crisis, and an uncertain future for all those I held dear, the moths had been a daily solace. Rays of light fluttering in the dark. They were a salve that—for reasons you now know—London had in much shorter supply.

Even in London, though, a moth trap has the capacity to surprise, and on July 17 mine did just that. The nighttime temperature was up a bit compared to the previous few days, but not markedly so. Yet moth numbers in the trap that morning were double those of the day before, and treble the day before that. I'd caught my first two terrace Jersey Tigers of 2020 on the sixteenth, but their numbers leapt to thirty-two on the seventeenth. Micromoths were also out in force. Horse Chestnut and Apple Leaf Miner numbers were both up, and I had my first Brown House Moths of the year. Plus, there was a micro I knew I hadn't seen before.

Catching a new species of moth always triggers a burst of adrenalin. There's a nervous fumbling with the camera to ensure a record shot or two, checking the settings are correct and the photographs in focus— not guaranteed for such small subjects. Then a gentle easing of the moth into a collecting tube, and into the fridge to chill in case the photo doesn't catch the necessary identification features. That's never guaranteed with micromoths, anyway (or even all macros), as many can only be named through recourse to dissection, currently beyond my abilities. This one looked possible, though. The prominent upturned "nose" formed by its palps, the flat elongated shape, and the parallel black streaks on a pale gray background, were a perfect match for one of the photographs in Chris Manley's identification guide. *Anarsia lineatella*, the Peach Twig Borer. Adrenalin leached into the warm glow that accompanies finding something new.

Peach Twig Borer is another moth not native to the UK, like the Box-tree, and so of particular interest to someone whose work is studying such species. I went online to find more on its back story. And that's when the Borer really turned interesting. The photos didn't match. Peach Twig Borer has more of a black spot in the middle of the wing, where my moth had a strong central streak. My moth was not a Peach Twig Borer at all—and in fact neither was the photograph in Manley. Most individuals coming to light identified as this species are something different, I discovered. A close relative, but still different. *Anarsia innoxiella*, the Acer Sober.

The real surprise, though, was that before 2017, the Acer Sober didn't even officially exist.

As its English name implies, *Anarsia lineatella* develops on fruit trees, where it's a pest of peach, plums, and other species in the genus *Prunus*. When it was first discovered in Denmark in the 1960s, the Danish State Plant Pathology Institute considered destroying all *Prunus* trees in the immediate vicinity in order to prevent its establishment on these economically important crops. It was lucky they didn't, because with hindsight, it was unnecessary. Specialists in the family to which *Anarsia* belongs had assumed for a while that moths identified as *lineatella* actually belonged to more than one species, but it was only in 2017 that taxonomists Keld Gregersen and Ole Karsholt showed that that was indeed the case. The Danish moths were *A. innoxiella*, which as their English name attests, develop on maples. And as their scientific name emphasizes, they are innocuous. The photograph in Manley was captioned correctly at the time of publication, just not in hindsight.

The buzz I get from identifying a species I've never seen before is something that I know will never leave me. The challenge presented, the pleasure of working towards an answer, the satisfaction of hitting the name on the head. All the greater that summer day in 2020, because the

Acer Sober had not even *had* a name when I first set a trap in Kindrogan, just four years earlier. It was a timely reminder that magic is never far away when you have a moth trap. It was a reminder of more than that, though.

I'd only been trying to name moths systematically for a couple of years, but in the process had learnt so much more than their identities. I'd opened a window into a whole new world of wonder through the lives of moths, and that had set me thinking about how the whole world is squeezed into the contents of the trap. But all the way I was reliant on the work of the many thousands of scientists and natural historians in whose footsteps I was walking. Without them, I would have been lost in the dark. We all would.

And finally, my first Acer Sober that July day was a reminder of how much there is still out there to learn. We think of the biodiversity of the temperate northern latitudes as well characterized by science, and indeed it is. But we also think that more than three-quarters of all animal species currently await a name. The vast majority of them reside in the tropics, where the numbers and density of taxonomists and scientists are low. They live in natural areas that are currently in full-scale retreat under the onslaught of chainsaw and plow. Many of those millions of species will disappear for good before we have even properly met them. No one will ever experience the thrill of identifying them for the first time. Their role in the story of life will never be told.

Yet everyone, everywhere, has such species living quietly around them, unseen, even in the hearts of our largest cities. With a moth trap, you can draw some of them into the light. Their presence is a product of the rules of nature and the cold hand of chance. These pressures have created jewels: Emeralds, Pearls, Rubies, and Gems. Every one of them is beautiful. Every one matters. Our lives will be better if we look out for them.

Sources

Books

Berryman, A. (Ed.). (2002). *Population Cycles*. Oxford University Press.

Forbush, E. H., & Fernald, C.H. (1896). *The Gypsy Moth. PORTHETRIA DISPAR (LINN.). A Report of the Work of destroying the insect in the commonwealth of Massachusetts, together with an Account of its History and Habits both in Massachusetts and Europe*. Wright & Potter Printing Co.

Gotelli, N. J. (1995). *A Primer of Ecology*. Sinauer Associates Inc.

Hanski, I. (1999). *Metapopulation Ecology*. Oxford University Press.

Hird, V. (2021). *Rebugging the Planet. The Remarkable Things that Insects (and Other Invertebrates) Do—and Why We Need to Love Them More*. Chelsea Green Publishing.

Hubbell, S. P. (2001). *The Unified Neutral Theory of Biodiversity and Biogeography*. Princeton University Press.

IPBES (2019). *Global assessment report of the Intergovernmental Science-Policy Platform on Biodiversity and Ecosystem Services*. (E. S. Brondízio, J. Settele, S. Díaz, & H. T. Ngo, Eds.). IPBES secretariat, Bonn, Germany.

Krebs, C. J. (2001). *Ecology* (5th ed.). Benjamin Cummings.

Lees, D. C., & Zilli, A. (2019). *Moths: Their Biology, Diversity and Evolution*. The Natural History Museum.

Leibold, M. A., & Chase, J. M. (2018). *Metacommunity Ecology*. Princeton University Press.

Lowen, J. (2021). *Much Ado about Mothing: A year intoxicated by Britain's rare and remarkable moths*. Bloomsbury Wildlife.

MacArthur, R. H., & Wilson, E. O. (1967). *The Theory of Island Biogeography*. Princeton University Press.

Majerus, M. (2002). *Moths*. Harper Collins.

Manley, C. (2015). *British Moths* (2nd ed.). Bloomsbury.

Marren, P. (2019). *Emperors, Admirals, and Chimney-Sweepers: The weird and wonderful names of butterflies and moths*. Little Toller Books.

Townsend, C. R., Begon, M., & Harper, J. L. (2003). *Essentials of Ecology* (2nd ed.). Blackwell Publishing.

Vellend, M. (2016). *The Theory of Ecological Communities*. Princeton University Press.

Waring, P., Townsend, M., & Lewington, R. (2009). *Field Guide to the Moths of Great Britain and Ireland* (2nd ed.). British Wildlife Publishing Ltd.

Whittaker, R. J. (1998). *Island Biogeography: Ecology, Evolution and Conservation*. Oxford University Press.

Whittaker, R. J., & Fernandez-Palacios, J. M. (2006). *Island Biogeography: Ecology, Evolution, and Conservation* (2nd ed.). Oxford University Press.

WWF (2020). *Living Planet Report 2020—Bending the curve of biodiversity loss*. (R. E. A. Almond, M. Grooten, & T. Petersen, Eds.). WWF, Gland, Switzerland.

Scientific Papers

Abang, F., & Karim, C. (2005). Diversity of macromoths (Lepidoptera: Heterocera) in the Poring Hill Dipterocarp Forest, Sabah, Borneo. *Journal of Asia-Pacific Entomology, 8,* 69–79.

Adler, P. B., HilleRisLambers, J., & Levine, J. M. (2007). A niche for neutrality. *Ecology Letters, 10,* 95–104.

Aguiar, A. P., Deans, A. R., Engel, M. S., Forshage, M., Huber, J. T., Jennings, J. T., Johnson, N. F., Lelej, A. S., Longino, J. T., Lohrmann, V., Mikó, I., Ohl, M., Rasmussen, C., Taeger, A., & Yu, D. S. K. (2013). Order Hymenoptera. *Zootaxa, 3703,* 51.

Ameca y Juárez, E. I., Mace, G. M., Cowlishaw, G., & Pettorelli, N. (2012). Natural population die-offs: Causes and consequences for terrestrial mammals. *Trends in Ecology & Evolution, 27,* 272–77.

Andersen, J. C., Havill, N. P., Griffin, B. P., Jepsen, J. U., Hagen, S. B., Klemola, T., Barrio, I. C., Kjeldgaard, S. A., Høye, T. T., Murlis, J., Baranchikov, Y. N., Selikhovkin, A. V., Vindstad, O. P. L., Caccone, A., & Elkinton, J. S. (2020). Northern Fennoscandia via the British Isles: Evidence for a novel post-glacial recolonization route by winter moth (*Operophtera brumata*). *Frontiers of Biogeography, 13,* e49581e.

Anderson, R. R., & May, R. M. (1980). Infectious diseases and population cycles of forest insects. *Science, 210,* 658–61.

Anon. (2011). Microbiology by numbers. *Nature Reviews Microbiology, 9,* 628.

Antão, L. H., Pöyry, J., Leinonen, R., & Roslin, T. (2020). Contrasting latitudinal patterns in diversity and stability in a high-latitude species-rich moth community. *Global Ecology and Biogeography, 29,* 896–907.

Baker, R. R. (1985). Moths: Population estimates, light-traps and migration. In L. M. Cook (Ed.). *Case Studies in Population Biology* (pp. 188–211). Manchester University Press.

Bakewell, A. T., Davis, K. E., Freckleton, R. P., Isaac, N. J. B., & Mayhew, P. J. (2020). Comparing life histories across taxonomic groups in multiple dimensions: How mammal-like are insects? *The American Naturalist, 195,* 70–81.

Ballesteros-Mejia, L., Kitching, I. J., Jetz, W., & Beck, J. (2017). Putting insects on the map: Near-global variation in Sphingid moth richness along spatial and environmental gradients. *Ecography, 40,* 698–708.

Baltensweiler, W. (1993). Why the larch bud-moth cycle collapsed in the subalpine larch-cembran pine forests in the year 1990 for the first time since 1850. *Oecologia, 94,* 62–66.

Barber, J., Plotkin, D., Rubin, J., Homziak, N., Leavell, B., Houlihan, P., Miner, K., Breinholt, J., Quirk-Royal, B., Padrón, P., Nunez, M., & Kawahara, A. (2021). Anti-bat ultrasound production in moths is globally and phylogenetically widespread. https://www.biorxiv.org/content/10.1101/2021.09.20.460855v1.full.pdf.

Bärtschi, F., McCain, C. M., Ballesteros-Mejia, L., Kitching, I. J., Beerli, N., & Beck, J. (2019). Elevational richness patterns of Sphingid moths support area effects over climatic drivers in a near-global analysis. *Global Ecology and Biogeography, 28,* 917–27.

Bates, A. J., Sadler, J. P., Grundy, D., Lowe, N., Davis, G., Baker, D., Bridge, M.,

Freestone, R., Gardner, D., Gibson, C., Hemming, R., Howarth, S., Orridge, S., Shaw, M., Tams, T., & Young, H. (2014). Garden and landscape-scale correlates of moths of differing conservation status: significant effects of urbanization and habitat diversity. *PLoS ONE, 9*, e86925.

Battin, J. (2004). When good animals love bad habitats: ecological traps and the conservation of animal populations. *Conservation Biology, 18*, 1482–91.

Beck, J., Kitching, I. J., & Linsenmair, K. E. (2006). Determinants of regional species richness: an empirical analysis of the number of hawkmoth species (Lepidoptera: Sphingidae) on the Malesian archipelago. *Journal of Biogeography, 33*, 694–706.

Beck, J., McCain, C. M., Axmacher, J. C., Ashton, L. A., Bärtschi, F., Brehm, G., Choi, S., Cizek, O., Colwell, R. K., Fiedler, K., Francois, C. L., Highland, S., Holloway, J. D., Intachat, J., Kadlec, T., Kitching, R. L., Maunsell, S. C., Merckx, T., Nakamura, A., Odell, E., Sang, W., Toko, P. S., Zamecnik, J., Zou, Y., Novotny, V., & Grytnes, J. (2017). Elevational species richness gradients in a hyperdiverse insect taxon: a global meta-study on geometrid moths. *Global Ecology and Biogeography, 26*, 412–24.

Beerli, N., Bärtschi, F., Ballesteros-Mejia, L., Kitching, I. J., & Beck, J. (2019). How has the environment shaped geographical patterns of insect body sizes? A test of hypotheses using Sphingid moths. *Journal of Biogeography, 46*, 1687–98.

Bell, J. R., Blumgart, D., & Shortall, C. R. (2020). Are insects declining and at what rate? An analysis of standardised, systematic catches of aphid and moth abundances across Great Britain. *Insect Conservation and Diversity, 13*, 115–26.

Belmaker, J., & Jetz, W. (2015). Relative roles of ecological and energetic constraints, diversification rates and region history on global species richness gradients. *Ecology Letters, 18*, 563–71.

Bethenod, M.-T., Thomas, Y., Rousset, F., Frérot, B., Pélozuelo, L., Genestier, G., & Bourguet, D. (2005). Genetic isolation between two sympatric host plant races of the European corn borer, *Ostrinia nubilalis* Hübner. II: Assortative mating and host-plant preferences for oviposition. *Heredity, 94*, 264–70.

Betzholtz, P.-E., Franzén, M., & Forsman, A. (2017). Colour pattern variation can inform about extinction risk in moths. *Animal Conservation, 20*, 72–79.

Bjornstad, O. N., Peltonen, M., Liebhold, A. M., & Baltensweiler, W. (2002). Waves of Larch Budmoth outbreaks in the European Alps. *Science, 298*, 1020–23.

Blackburn, T. M., Bellard, C., & Ricciardi, A. (2019). Alien versus native species as drivers of recent extinctions. *Frontiers in Ecology and the Environment, 17*, 203–7.

Blumgart, D., Botham, M. S., Menéndez, R., & Bell, J. R. (2022). Moth declines are most severe in broadleaf woodlands despite a net gain in habitat availability. *Insect Conservation and Diversity, 15*, 496–509.

Bonsall, M. B., & Hassell, M. P. (1997). Apparent competition structures ecological assemblages. *Nature, 388*, 371–73.

Boyes, D. H., Evans, D. M., Fox, R., Parsons, M. S., & Pocock, M. J. O. (2021a). Is light pollution driving moth population declines? A review of causal mechanisms across the life cycle. *Insect Conservation and Diversity, 14*, 167–87.

Boyes, D. H., Evans, D. M., Fox, R., Parsons, M. S., & Pocock, M. J. O. (2021b). Street lighting has detrimental impacts on local insect populations. *Science Advances, 7*, eabi8322.

Boyes, D. H., & Lewis, O. T. (2019). Ecology of Lepidoptera associated with bird nests in mid-Wales, UK. *Ecological Entomology, 44*, 1–10.

Broad, G. R., & Shaw, M. R. (2016). The British species of *Enicospilus* (Hymenoptera: Ichneumonidae: Ophioninae). *European Journal of Taxonomy, 187*, 1–31.

Bruzzese, D. J., Wagner, D. L., Harrison, T., Jogesh, T., Overson, R. P., Wickett, N. J., Raguso, R. A., & Skogen, K. A. (2019). Phylogeny, host use, and diversification in the moth family Momphidae (Lepidoptera: Gelechioidea). *PLOS ONE, 14*, e0207833.

Bull, J. W., & Maron, M. (2016). How humans drive speciation as well as extinction. *Proceedings of the Royal Society B: Biological Sciences, 283*, 20160600.

Büntgen, U., Liebhold, A., Nievergelt, D., Wermelinger, B., Roques, A., Reinig, F., Krusic, P. J., Piermattei, A., Egli, S., Cherubini, P., & Esper, J. (2020). Return of the moth: Rethinking the effect of climate on insect outbreaks. *Oecologia, 192*, 543–52.

Burner, R. C., Selås, V., Kobro, S., Jacobsen, R. M., & Sverdrup-Thygeson, A. (2021). Moth species richness and diversity decline in a 30-year time series in Norway, irrespective of species' latitudinal range extent and habitat. *Journal of Insect Conservation, 25*, 887–96.

Burns, F., Eaton, M. A., Burfield, I. J., Klvaňová, A., Šilarová, E., Staneva, A., & Gregory, R. D. (2021). Abundance decline in the avifauna of the European Union reveals cross-continental similarities in biodiversity change. *Ecology and Evolution, 11*, 16647–60.

Canfield, M. R., Greene, E., Moreau, C. S., Chen, N., & Pierce, N. E. (2008). Exploring phenotypic plasticity and biogeography in emerald moths: A phylogeny of the genus *Nemoria* (Lepidoptera: Geometridae). *Molecular Phylogenetics and Evolution, 49*, 477–87.

Cannon, P. G., Edwards, D. P., & Freckleton, R. P. (2021). Asking the wrong question in explaining tropical diversity. *Trends in Ecology & Evolution, 36*, 482–84.

Carde, R. T. Insect migration: Do migrant moths know where they are heading? *Current Biology, 18*, R472–74.

Carr, A., Weatherall, A., Fialas, P., Zeale, M. R. K., Clare, E. L., & Jones, G. (2020). Moths consumed by the Barbastelle *Barbastella barbastellus* require larval host plants that occur within the bat's foraging habitats. *Acta Chiropterologica, 22*, 257–69.

Catford, J. A., Bode, M., & Tilman, D. (2018). Introduced species that overcome life history trade-offs can cause native extinctions. *Nature Communications, 9*, 2131.

Chao, A., & Chiu, C. (2016). Species richness: estimation and comparison. In N. Balakrishnan, T. Colton, B. Everitt, W. Piegorsch, F. Ruggeri, & J. L. Teugels (Eds.), *Wiley StatsRef: Statistics Reference Online* (pp. 1–26). Wiley.

Chesson, P. (2000). Mechanisms of maintenance of species diversity. *Annual Review of Ecology and Systematics, 31*, 343–66.

Chown, S. L., & Gaston, K. J. (2010). Body size variation in insects: A macroecological perspective. *Biological Reviews, 85*, 139–69.

Church, S. H., Donoughe, S., de Medeiros, B. A. S., & Extavour, C. G. (2019). Insect egg size and shape evolve with ecology but not developmental rate. *Nature, 571*, 58–62.

Cole, E. F., Regan, C. E., & Sheldon, B. C. (2021). Spatial variation in avian phenological response to climate change linked to tree health. *Nature Climate Change, 11*, 872–78.

Conrad, K. F., Warren, M. S., Fox, R., Parsons, M. S., & Woiwod, I. P. (2006). Rapid declines of common, widespread British moths provide evidence of an insect biodiversity crisis. *Biological Conservation, 132,* 279–91.

Cook, L. M., & Graham, C. S. (1996). Evenness and species number in some moth populations. *Biological Journal of the Linnean Society, 58,* 75–84.

Correa-Carmona, Y., Rougerie, R., Arnal, P., Ballesteros-Mejia, L., Beck, J., Dolédec, S., Ho, C., Kitching, I. J., Lavelle, P., Le Clec'h, S., Lopez-Vaamonde, C., Martins, M. B., Murienne, J., Oszwald, J., Ratnasingham, S., & Decaëns, T. (2022). Functional and taxonomic responses of tropical moth communities to deforestation. *Insect Conservation and Diversity, 15,* 236–47.

Crawley, M. J., & Pattrasudhi, R. (1988). Interspecific competition between insect herbivores: asymmetric competition between cinnabar moth and the ragwort seed-head fly. *Ecological Entomology, 13,* 243–49.

Crawley, M. J., & Gillman, M. P. (1989). Population dynamics of Cinnabar Moth and Ragwort in grassland. *Journal of Animal Ecology, 58,* 1035–50.

Crouch, N. M. A., & Tobias, J. A. (2022). The causes and ecological context of rapid morphological evolution in birds. *Ecology Letters, 25,* 611–23.

Danks, H. V. (1992). Long life cycles in insects. *The Canadian Entomologist, 124,* 167–87.

Dapporto, L., & Dennis, R. L. H. (2013). The generalist–specialist continuum: Testing predictions for distribution and trends in British butterflies. *Biological Conservation, 157,* 229–36.

Darimont, C. T., Carlson, S. M., Kinnison, M. T., Paquet, P. C., Reimchen, T. E., & Wilmers, C. C. (2009). Human predators outpace other agents of trait change in the wild. *Proceedings of the National Academy of Sciences, USA, 106,* 952–54.

Davies, T. J. (2021). Ecophylogenetics redux. *Ecology Letters, 24,* 1073–88.

Davis, R. B., Javoiš, J., Pienaar, J., & Unap, E. O. (2012). Disentangling determinants of egg size in the Geometridae (Lepidoptera) using an advanced phylogenetic comparative method. *Journal of Evolutionary Biology, 25,* 210–19.

Dempster, J. P. (1983). The natural control of populations of butterflies and moths. *Biological Reviews, 58,* 461–81.

Dennis, E. B., Morgan, B. J. T., Freeman, S. N., Brereton, T. M., & Roy, D. B. (2016). A generalized abundance index for seasonal invertebrates. *Biometrics, 72,* 1305–14.

Denno, R. F., McClure, M. S., & Ott, J. R. (1995). Interspecific interactions in phytophagous insects: competition reexamined and resurrected. *Annual Review of Entomology, 40,* 297–331.

Diaz, R. M., Ye, H., & Ernest, S. K. M. (2021). Empirical abundance distributions are more uneven than expected given their statistical baseline. *Ecology Letters, 24,* 2025–39.

Drury, J. P., Clavel, J., Tobias, J. A., Rolland, J., Sheard, C., & Morlon, H. (2021). Tempo and mode of morphological evolution are decoupled from latitude in birds. *PLOS Biology, 19,* e3001270.

Ehlers, B. K., Bataillon, T., & Damgaard, C. F. (2021). Ongoing decline in insect-pollinated plants across Danish grasslands. *Biology Letters, 17,* 20210493.

Elkinton, J. S., & Liebhold, A. M. (1990). Population dynamics of Gypsy Moth in North America. *Annual Review of Entomology, 35,* 571–96.

Elliott, C. H., Gillett, C. P.D. T., Parsons, E., Wright, M. G., & Rubinoff, D. (2022).

Identifying key threats to a refugial population of an endangered Hawaiian moth. *Insect Conservation and Diversity, 15,* 263–72.

Ellis, E. E., & Wilkinson, T. L. (2020). Moth assemblages within urban domestic gardens respond positively to habitat complexity, but only at a scale that extends beyond the garden boundary. *Urban Ecosystems, 24,* 469–79.

van Els, P., Herrera-Alsina, L., Pigot, A. L., & Etienne, R. S. (2021). Evolutionary dynamics of the elevational diversity gradient in passerine birds. *Nature Ecology & Evolution, 5,* 1259–65.

Elton, C., & Nicholson, M. (1942). The ten-year cycle in numbers of the lynx in Canada. *Journal of Animal Ecology, 11,* 215–44.

Farrell, B. D., Mitter, C., & Futuyma, D. J. (1992). Diversification at the Insect-Plant Interface. *BioScience, 42,* 34–42.

Feeny, P. (1970). Seasonal changes in oak leaf tannins and nutrients as a cause of spring feeding by winter moth caterpillars. *Ecology, 51,* 565–81.

Fenoglio, M. S., Calviño, A., González, E., Salvo, A., & Videla, M. (2021). Urbanisation drivers and underlying mechanisms of terrestrial insect diversity loss in cities. *Ecological Entomology, 46,* 757–71.

Fine, P. V. A. (2015). Ecological and evolutionary drivers of geographic variation in species diversity. *Annual Review of Ecology, Evolution, and Systematics, 46,* 369–92.

Fisher, K. (1938). Migrations of the Silver-Y Moth (*Plusia gamma*) in Great Britain. *Journal of Animal Ecology, 7,* 230–47.

Fisher, R. A., Corbet, A. S., & Williams, C. B. (1943). The relation between the number of species and the number of individuals in a random sample of an animal population. *Journal of Animal Ecology, 12,* 42–58.

Forbes, A. A., Bagley, R. K., Beer, M. A., Hippee, A. C., & Widmayer, H. A. (2018). Quantifying the unquantifiable: Why Hymenoptera, not Coleoptera, is the most speciose animal order. *BMC Ecology, 18,* 21.

Forsman, A., Betzholtz, P.-E., & Franzén, M. (2016). Faster poleward range shifts in moths with more variable colour patterns. *Scientific Reports, 6,* 36265.

Forsman, A., Betzholtz, P.-E., & Franzén, M. (2015). Variable coloration is associated with dampened population fluctuations in noctuid moths. *Proceedings of the Royal Society B: Biological Sciences, 282,* 20142922.

Forsman, A., Polic, D., Sunde, J., Betzholtz, P., & Franzén, M. (2020). Variable colour patterns indicate multidimensional, intraspecific trait variation and ecological generalization in moths. *Ecography, 43,* 823–33.

Fourcade, Y., WallisDeVries, M. F., Kuussaari, M., Swaay, C. A. M., Heliölä, J., & Öckinger, E. (2021). Habitat amount and distribution modify community dynamics under climate change. *Ecology Letters, 24,* 950–57.

Fox, R. (2013). The decline of moths in Great Britain: A review of possible causes. *Insect Conservation and Diversity, 6,* 5–19.

Fox, R., Oliver, T. H., Harrower, C., Parsons, M. S., Thomas, C. D., & Roy, D. B. (2014). Long-term changes to the frequency of occurrence of British moths are consistent with opposing and synergistic effects of climate and land-use changes. *Journal of Applied Ecology, 51,* 949–57.

Fox, R., Randle, Z., Hill, L., Anders, S., Wiffen, L., & Parsons, M. S. (2011). Moths count: Recording moths for conservation in the UK. *Journal of Insect Conservation*, *15*, 55–68.

Fragata, I., Costa-Pereira, R., Kozak, M., Majer, A., Godoy, O., & Magalhães, S. (2022). Specific sequence of arrival promotes coexistence via spatial niche pre-emption by the weak competitor. *Ecology Letters, 25*, 1629–39.

Franzén, M., Betzholtz, P.-E., Pettersson, L. B., & Forsman, A. (2020). Urban moth communities suggest that life in the city favours thermophilic multi-dimensional generalists. *Proceedings of the Royal Society B: Biological Sciences, 287*, 20193014.

Franzén, M., Forsman, A., & Betzholtz, P. (2019). Variable color patterns influence continental range size and species–area relationships on islands. *Ecosphere, 10*, e02577.

Franzén, M., Schweiger, O., & Betzholtz, P.-E. (2012). Species-area relationships are controlled by species traits. *PLoS ONE, 7*, e37359.

Fraser, S. M., & Lawton, J. H. (1994). Host range expansion by British moths onto introduced conifers. *Ecological Entomology, 19*, 127–37.

Fretwell, S. D. (1975). The impact of Robert MacArthur on ecology. *Annual Review of Ecology and Systematics, 6*, 1–13.

Fuentes-Montemayor, E., Goulson, D., Cavin, L., Wallace, J. M., & Park, K. J. (2012). Factors influencing moth assemblages in woodland fragments on farmland: Implications for woodland management and creation schemes. *Biological Conservation, 153*, 265–75.

García-Barros, E. (2000). Body size, egg size, and their interspecific relationships with ecological and life history traits in butterflies (Lepidoptera: Papilionoidea, Hesperioidea). *Biological Journal of the Linnean Society, 70*, 251–84.

Gaston, K. J. (1988). Patterns in the local and regional dynamics of moth populations. *Oikos, 53*, 49–57.

Geffen, K. G., Grunsven, R. H. A., Ruijven, J., Berendse, F., & Veenendaal, E. M. (2014). Artificial light at night causes diapause inhibition and sex-specific life history changes in a moth. *Ecology and Evolution, 4*, 2082–89.

Gilbert, J. D. J., & Manica, A. (2010). Parental care trade-offs and life-history relationships in insects. *The American Naturalist, 176*, 212–26.

Gilioli, G., Bodini, A., Cocco, A., Lentini, A., & Luciano, P. (2012). Analysis and modelling of *Lymantria dispar* (L.) metapopulation dynamics in Sardinia. *OBC/wprs Bulletin, 76*, 163–70.

Godfray, H. C. J., Partridge, L., & Harvey, P. H. (1991). Clutch size. *Annual Review of Ecology and Systematics, 2*, 409–29.

Goldstein, P. Z., Morita, S., & Capshaw, G. (2015). Stasis and flux among Saturniidae and Sphingidae (Lepidoptera) on Massachusetts' offshore islands and the possible role of *Compsilura concinnata* (Meigen) (Diptera: Tachinidae) as an agent of mainland New England moth declines. *Proceedings of the Entomological Society of Washington, 117*, 347–66.

Gooriah, L., Blowes, S. A., Sagouis, A., Schrader, J., Karger, D. N., Kreft, H., & Chase, J. M. (2021). Synthesis reveals that island species–area relationships emerge from processes beyond passive sampling. *Global Ecology and Biogeography, 30*, 2119–31.

Gotelli, N. J., & Kelley, W. G. (1993). A general model of metapopulation dynamics. *Oikos, 68*, 36.

Gregersen, K., & Karsholt, O. (2017). Taxonomic confusion around the Peach Twig Borer, *Anarsia lineatella* Zeller, 1839, with description of a new species (Lepidoptera, Gelechiidae). *Nota Lepidopterologica, 40,* 65–85.

Grenyer, R., Orme, C. D. L., Jackson, S. F., Thomas, G. H., Davies, R. G., Davies, T. J., Jones, K. E., Olson, V. A., Ridgely, R. S., Rasmussen, P. C., Ding, T.-S., Bennett, P. M., Blackburn, T. M., Gaston, K. J., Gittleman, J. L., & Owens, I. P. F. (2006). Global distribution and conservation of rare and threatened vertebrates. *Nature, 444,* 93–96.

Gripenberg, S., Ovaskainen, O., Morriën, E., & Roslin, T. (2008). Spatial population structure of a specialist leaf-mining moth. *Journal of Animal Ecology, 77,* 757–67.

Grünig, M., Beerli, N., Ballesteros-Mejia, L., Kitching, I. J., & Beck, J. (2017). How climatic variability is linked to the spatial distribution of range sizes: seasonality versus climate change velocity in sphingid moths. *Journal of Biogeography, 44,* 2441–50.

van Grunsven, R. H. A., van Deijk, J. R., Donners, M., Berendse, F., Visser, M. E., Veenendaal, E., & Spoelstra, K. (2020). Experimental light at night has a negative long-term impact on macro-moth populations. *Current Biology, 30,* R694–95.

Hahn, M., Schotthöfer, A., Schmitz, J., Franke, L. A., & Brühl, C. A. (2015). The effects of agrochemicals on Lepidoptera, with a focus on moths, and their pollination service in field margin habitats. *Agriculture, Ecosystems & Environment, 207,* 153–62.

Hanski, I., & Gyllenberg, M. (1997). Uniting two general patterns in the distribution of species. *Science, 275,* 397–400.

Harmon, L. J., & Harrison, S. (2015). Species diversity is dynamic and unbounded at local and continental scales. *The American Naturalist, 185,* 584–93.

Harrison, S., & Karban, R. (1986). Effects of an early-season folivorous moth on the success of a later-season species, mediated by a change in the quality of the shared host, *Lupinus arboreus* Sims. *Oecologia, 69,* 354–59.

Harrower, C. A., Bell, J. R., Blumgart, D., Botham, M. S., Fox, R., Isaac, N. J. B., Roy, D. B., & Shortall, C. R. (2020). Moth trends for Britain and Ireland from the Rothamsted Insect Survey light-trap network (1968 to 2016). NERC Environmental Information Data Centre. (Dataset). https://doi.org/10.5285/0a7d65e8-8bc8-46e5-ab72-ee64ed851583.

Hassell, M. P. (1975). Density-dependence in single-species populations. *Journal of Animal Ecology, 44,* 283–95.

Hassell, M. P., Crawley, M. J., Godfray, H. C. J., & Lawton, J. H. (1998). Top-down versus bottom-up and the Ruritanian bean bug. *Proceedings of the National Academy of Sciences, USA, 95,* 10661–64.

Healy, K., Ezard, T. H. G., Jones, O. R., Salguero-Gómez, R., & Buckley, Y. M. (2019). Animal life history is shaped by the pace of life and the distribution of age-specific mortality and reproduction. *Nature Ecology & Evolution, 3,* 1217–24.

Heidrich, L., Pinkert, S., Brandl, R., Bässler, C., Hacker, H., Roth, N., Busse, A., Müller, J., & Friess, N. (2021). Noctuid and geometrid moth assemblages show divergent elevational gradients in body size and color lightness. *Ecography, 44,* 1169–79.

Hembry, D. H., Bennett, G., Bess, E., Cooper, I., Jordan, S., Liebherr, J., Magnacca, K. N., Percy, D. M., Polhemus, D. A., Rubinoff, D., Shaw, K. L., & O'Grady, P. M. (2021). Insect radiations on islands: biogeographic pattern and evolutionary process in Hawaiian insects. *The Quarterly Review of Biology, 96,* 247–96.

Hendry, A. P., Gotanda, K. M., & Svensson, E. I. (2017). Human influences on evolution, and the ecological and societal consequences. *Philosophical Transactions of the Royal Society B: Biological Sciences, 372*, 20160028.

Hill, G. M., Kawahara, A. Y., Daniels, J. C., Bateman, C. C., & Scheffers, B. R. (2021). Climate change effects on animal ecology: butterflies and moths as a case study. *Biological Reviews, 96*, 2113–26.

Hoffmann, J. H., Moran, V. C., Zimmermann, H. G., & Impson, F. A. C. (2020). Biocontrol of a prickly pear cactus in South Africa: Reinterpreting the analogous, renowned case in Australia. *Journal of Applied Ecology, 57*, 2475–84.

ter Hofstede, H. M., & Ratcliffe, J. M. (2016). Evolutionary escalation: The bat–moth arms race. *Journal of Experimental Biology, 219*, 1589–602.

Holland, R. A., Wikelski, M., & Wilcove, D. S. (2006). How and why do insects migrate? *Science, 313*, 794–96.

Holm, S., Davis, R. B., Javoiš, J., Õunap, E., Kaasik, A., Molleman, F., & Tammaru, T. (2016). A comparative perspective on longevity: The effect of body size dominates over ecology in moths. *Journal of Evolutionary Biology, 29*, 2422–35.

Holm, S., Javoiš, J., Kaasik, A., Õunap, E., Davis, R. B., Molleman, F., Roininen, H., & Tammaru, T. (2019). Size-related life-history traits in geometrid moths: A comparison of a temperate and a tropical community. *Ecological Entomology, 44*, 711–16.

Hu, G., Lim, K. S., Horvitz, N., Clark, S. J., Reynolds, D. R., Sapir, N., & Chapman, J. W. (2016). Mass seasonal bioflows of high-flying insect migrants. *Science, 354*, 1584–87.

Hughes, A. C., Orr, M. C., Ma, K., Costello, M. J., Waller, J., Provoost, P., Yang, Q., Zhu, C., & Qiao, H. (2021). Sampling biases shape our view of the natural world. *Ecography, 44*, 1259–69.

Hughes, E. C., Edwards, D. P., Bright, J. A., Capp, E. J. R., Cooney, C. R., Varley, Z. K., & Thomas, G. H. (2022). Global biogeographic patterns of avian morphological diversity. *Ecology Letters, 25*, 598–610.

Hunter, M. D., Varley, G. C., & Gradwell, G. R. (1997). Estimating the relative roles of top-down and bottom-up forces on insect herbivore populations: A classic study revisited. *Proceedings of the National Academy of Sciences, USA, 94*, 9176–81.

Hunter, M. D., & Willmer, P. G. (1989). The potential for interspecific competition between two abundant defoliators on oak: leaf damage and habitat quality. *Ecological Entomology, 14*, 267–77.

Inkinen, P. (1994). Distribution and abundance in British noctuid moths revisited. *Annales Zoologici Fennici, 31*, 235–43.

Isaac, N. J. B., Jones, K. E., Gittleman, J. L., & Purvis, A. (2005). Correlates of species richness in mammals: Body size, life history, and ecology. *The American Naturalist, 165*, 600–7.

Jankovic, M., & Petrovskii, S. (2013). Gypsy moth invasion in North America: A simulation study of the spatial pattern and the rate of spread. *Ecological Complexity, 14*, 132–44.

Janzen, D. H. (1967). Why mountain passes are higher in the tropics. *The American Naturalist, 101*, 233–49.

Jarzyna, M. A., Quintero, I., & Jetz, W. (2021). Global functional and phylogenetic structure of avian assemblages across elevation and latitude. *Ecology Letters, 24,* 196–207.

Jepson, P. D., Deaville, R., Barber, J. L., Aguilar, À., Borrell, A., Murphy, S., Barry, J., Brownlow, A., Barnett, J., Berrow, S., Cunningham, A. A., Davison, N. J., ten Doeschate, M., Esteban, R., Ferreira, M., Foote, A. D., Genov, T., Giménez, J., Loveridge, J., Llavona, Á., Martin, V., Maxwell, D. L., Papachlimitzou, A., Penrose, R., Perkins, M. W., Smith, B., de Stephanis, R., Tregenza, N., Verborgh, P., Fernandez, A., & Law, R. J. (2016). PCB pollution continues to impact populations of orcas and other dolphins in European waters. *Scientific Reports, 6,* 18573.

Jervis, M. A., Boggs, C. L., & Ferns, P. N. (2007a). Egg maturation strategy and survival trade-offs in holometabolous insects: A comparative approach. *Biological Journal of the Linnean Society, 90,* 293–302.

Jervis, M. A., Ferns, P. N., & Boggs, C. L. (2007b). A trade-off between female lifespan and larval diet breadth at the interspecific level in Lepidoptera. *Evolutionary Ecology, 21,* 307–23.

Jetz, W., Thomas, G. H., Joy, J. B., Hartmann, K., & Mooers, A. O. (2012). The global diversity of birds in space and time. *Nature, 491,* 444–48.

Johnson, D. M., Liebhold, A. M., Tobin, P. C., & Bjørnstad, O. N. (2006). Allee effects and pulsed invasion by the gypsy moth. *Nature, 444,* 361–63.

Kawahara, A. Y., Plotkin, D., Espeland, M., Meusemann, K., Toussaint, E. F. A., Donath, A., Gimnich, F., Frandsen, P. B., Zwick, A., dos Reis, M., Barber, J. R., Peters, R. S., Liu, S., Zhou, X., Mayer, C., Podsiadlowski, L., Storer, C., Yack, J. E., Misof, B., & Breinholt, J. W. (2019). Phylogenomics reveals the evolutionary timing and pattern of butterflies and moths. *Proceedings of the National Academy of Sciences, USA, 116,* 22657–63.

Kawahara, A. Y., Reeves, L. E., Barber, J. R., & Black, S. H. (2021). Opinion: Eight simple actions that individuals can take to save insects from global declines. *Proceedings of the National Academy of Sciences, USA, 118,* e2002547117.

Kinsella, R. S., Thomas, C. D., Crawford, T. J., Hill, J. K., Mayhew, P. J., & Macgregor, C. J. (2020). Unlocking the potential of historical abundance datasets to study biomass change in flying insects. *Ecology and Evolution, 10,* 8394–404.

Kubelka, V., Sandercock, B. K., Székely, T., & Freckleton, R. P. (2021). Animal migration to northern latitudes: environmental changes and increasing threats. *Trends in Ecology & Evolution, 37,* 30–41.

Lamarre, G. P. A., Pardikes, N. A., Segar, S., Hackforth, C. N., Laguerre, M., Vincent, B., Lopez, Y., Perez, F., Bobadilla, R., Silva, J. A. R., & Basset, Y. (2022). More winners than losers over 12 years of monitoring tiger moths (Erebidae: Arctiinae) on Barro Colorado Island, Panama. *Biology Letters, 18,* 20210519.

Leibhold, A., Mastro, V., & Schaefer, P. W. (1989). Learning from the legacy of Léopold Trouvelot. *Bulletin of the Entomological Society of America, 35,* 20–22.

Li, H., & Wiens, J. J. (2019). Time explains regional richness patterns within clades more often than diversification rates or area. *The American Naturalist, 193,* 514–29.

Li, P., & Wiens, J. J. (2022). What drives diversification? Range expansion tops climate, life history, habitat and size in lizards and snakes. *Journal of Biogeography, 49,* 237–47.

Liebhold, A., Elkinton, J., Williams, D., & Muzika, R.-M. (2000). What causes outbreaks of the gypsy moth in North America? *Population Ecology, 42,* 257–66.

Liebhold, A. M., Halverson, J. A., & Elmes, G. A. (1992). Gypsy moth invasion in North America: A quantitative analysis. *Journal of Biogeography, 19*, 513.

Liebhold, A. M., Haynes, K. J., & Bjørnstad, O. N. (2012). Spatial synchrony of insect outbreaks. insect outbreaks revisited. In P. Barbosa, D. K. Letourneau, & A. A. Agrawal (Eds.). *Insect Outbreaks Revisited* (pp. 113–25). John Wiley & Sons, Ltd.

Carrière, Y., Deland, J. P., Roff, D. A., & Vincent, C. (1994). Life-history costs associated with the evolution of insecticide resistance. *Proceedings of the Royal Society of London. Series B: Biological Sciences, 258*, 35–40.

Lindström, J., Kaila, L., & Niemelä, P. (1994). Polyphagy and adult body size in geometrid moths. *Oecologia, 98*, 130–32.

Lintott, P. R., Bunnefeld, N., Fuentes-Montemayor, E., Minderman, J., Blackmore, L. M., Goulson, D., & Park, K. J. (2014). Moth species richness, abundance and diversity in fragmented urban woodlands: Implications for conservation and management strategies. *Biodiversity and Conservation, 23*, 2875–901.

Lockett, M. T., Jones, T. M., Elgar, M. A., Gaston, K. J., Visser, M. E., & Hopkins, G. R. (2021). Urban street lighting differentially affects community attributes of airborne and ground-dwelling invertebrate assemblages. *Journal of Applied Ecology, 58*, 2329–39.

Loder, N., Gaston, K. J., Warren, P. H., & Arnold, H. R. (1998). Body size and feeding specificity: Macrolepidoptera in Britain. *Biological Journal of the Linnean Society, 63*, 121–39.

Loss, S. R., Will, T., & Marra, P. P. (2013). The impact of free-ranging domestic cats on wildlife of the United States. *Nature Communications, 4*, 1396.

Macgregor, C. J., Williams, J. H., Bell, J. R., & Thomas, C. D. (2019). Moth biomass has fluctuated over 50 years in Britain but lacks a clear trend. *Nature Ecology & Evolution, 3*, 1645–49.

Macgregor, C. J., Williams, J. H., Bell, J. R., & Thomas, C. D. (2021). Author correction: Moth biomass has fluctuated over 50 years in Britain but lacks a clear trend. *Nature Ecology & Evolution, 5*, 865–83.

Machac, A., & Graham, C. H. (2017). Regional diversity and diversification in mammals. *The American Naturalist, 189*, E1–13.

Mahmoudvand, M., Abbasipour, H., Garjan, A. S., & Bandani, A. R. (2011). Sublethal effects of indoxacarb on the diamondback moth, *Plutella xylostella* (L.) (Lepidoptera: Yponomeutidae). *Applied Entomology and Zoology, 46*, 75–80.

Mally, R., Turner, R. M., Blake, R. E., Fenn, G., Bertelsmeier, C., Brockerhoff, E. G., Hoare, R. J. B., Nahrung, H. F., Roques, A., Pureswaran, D. S., Yamanaka, T., & Liebhold, A. M. (2022). Moths and butterflies on alien shores: Global biogeography of non-native Lepidoptera. *Journal of Biogeography, 49*, 1455–68.

Mannion, P. D., Upchurch, P., Benson, R. B. J., & Goswami, A. (2014). The latitudinal biodiversity gradient through deep time. *Trends in Ecology & Evolution, 29*, 42–50.

Martay, B., Brewer, M. J., Elston, D. A., Bell, J. R., Harrington, R., Brereton, T. M., Barlow, K. E., Botham, M. S., & Pearce-Higgins, J. W. (2017). Impacts of climate change on national biodiversity population trends. *Ecography, 40*, 1139–51.

Mason, S. C., Palmer, G., Fox, R., Gillings, S., Hill, J. K., Thomas, C. D., & Oliver, T. H. (2015). Geographical range margins of many taxonomic groups continue to shift polewards. *Biological Journal of the Linnean Society, 115*, 586–97.

Matthews, T. J. (2021). On the biogeography of habitat islands: The importance of matrix effects, noncore species, and source-sink dynamics. *Quarterly Review of Biology, 96,* 73–104.

Matthews, T. J., Rigal, F., Triantis, K. A., & Whittaker, R. J. (2019). A global model of island species–area relationships. *Proceedings of the National Academy of Sciences, USA, 116,* 12337–42.

Mayhew, P. J. (2007). Why are there so many insect species? Perspectives from fossils and phylogenies. *Biological Reviews, 82,* 425–54.

McManus, M., & Csóka, G. (2007). History and impact of Gypsy Moth in North America and comparison to recent outbreaks in Europe. *Acta Silvatica et Lignaria Hungarica, 3,* 47–64.

van der Meijden, E., & Wijk, C. van der V. (1997). Tritrophic metapopulation dynamics. A case study of Ragwort, the Cinnabar Moth, and the parasitoid *Cotesia popularis*. In I. K. Hanski & M. E. Gilpin (Eds.). *Metapopulaion Biology: Ecology, Genetics, and Evolution* (pp. 387–405). Academic Press, San Diego.

Menken, S. B. J., Boomsma, J. J., & Van Nieukerken, E. J. (2009). Large-scale evolutionary patterns of host plant associations in the Lepidoptera: Host plant use in the Lepidoptera. *Evolution, 64,* 1098–119.

Merckx, T., Dantas de Miranda, M., & Pereira, H. M. (2019). Habitat amount, not patch size and isolation, drives species richness of macro-moth communities in countryside landscapes. *Journal of Biogeography, 46,* 956–67.

Merckx, T., Marini, L., Feber, R. E., & Macdonald, D. W. (2012). Hedgerow trees and extended-width field margins enhance macro-moth diversity: Implications for management. *Journal of Applied Ecology, 49,* 1396–404.

Merckx, T., Nielsen, M. E., Heliölä, J., Kuussaari, M., Pettersson, L. B., Pöyry, J., Tiainen, J., Gotthard, K., & Kivelä, S. M. (2021). Urbanization extends flight phenology and leads to local adaptation of seasonal plasticity in Lepidoptera. *Proceedings of the National Academy of Sciences, USA, 118,* e2106006118.

Mitchell, A., Mitter, C., & Regier, J. C. (2005). Systematics and evolution of the cutworm moths (Lepidoptera: Noctuidae): Evidence from two protein-coding nuclear genes: Molecular systematics of Noctuidae. *Systematic Entomology, 31,* 21–46.

Mittelbach, G. G., Schemske, D. W., Cornell, H. V., Allen, A. P., Brown, J. M., Bush, M. B., Harrison, S. P., Hurlbert, A. H., Knowlton, N., Lessios, H. A., McCain, C. M., McCune, A. R., McDade, L. A., McPeek, M. A., Near, T. J., Price, T. D., Ricklefs, R. E., Roy, K., Sax, D. F., Schluter, D., Sobel, J. M., & Turelli, M. (2007). Evolution and the latitudinal diversity gradient: speciation, extinction and biogeography. *Ecology Letters, 10,* 315–31.

Mitter, C., Davis, D. R., & Cummings, M. P. (2017). Phylogeny and evolution of Lepidoptera. *Annual Review of Entomology, 62,* 265–83.

Mottl, O., Flantua, S. G. A., Bhatta, K. P., Felde, V. A., Giesecke, T., Goring, S., Grimm, E. C., Haberle, S., Hooghiemstra, H., Ivory, S., Kuneš, P., Wolters, S., Seddon, A. W. R., & Williams, J. W. (2021). Global acceleration in rates of vegetation change over the past 18,000 years. *Science, 372,* 860–64.

Mutshinda, C. M., O'Hara, R. B., & Woiwod, I. P. (2008). Species abundance dynamics under neutral assumptions: A Bayesian approach to the controversy. *Functional Ecology, 22,* 340–47.

Mutshinda, C. M., O'Hara, R. B., & Woiwod, I. P. (2009). What drives community dynamics? *Proceedings of the Royal Society B: Biological Sciences, 276*, 2923–29.

Nieminen, M. (1996). Risk of population extinction in moths: Effect of host plant characteristics. *Oikos, 76*, 475–84.

Nieminen, M., & Hanski, I. (1998). Metapopulations of moths on islands: A test of two contrasting models. *Journal of Animal Ecology, 67*, 149–60.

Nilsson, L. A. (1983). Processes of isolation and introgressive interplay between *Platanthera bifolia* (L.) Rich and *P. chlorantha* (Custer) Reichb. (Orchidaceae). *Botanical Journal of the Linnean Society, 87*, 325–50.

Nogué, S., Santos, A. M. C., Birks, H. J. B., Björck, S., Castilla-Beltrán, A., Connor, S., de Boer, E. J., de Nascimento, L., Felde, V. A., Fernández-Palacios, J. M., Froyd, C. A., Haberle, S. G., Hooghiemstra, H., Ljung, K., Norder, S. J., Peñuelas, J., Prebble, M., Stevenson, J., Whittaker, R. J., Willis, K. J., Wilmshurst, J. M., & Steinbauer, M. J. (2021). The human dimension of biodiversity changes on islands. *Science, 372*, 488–91.

Nunes, C. A., Berenguer, E., França, F., Ferreira, J., Lees, A. C., Louzada, J., Sayer, E. J., Solar, R., Smith, C. C., Aragão, L. E. O. C., Braga, D. de L., de Camargo, P. B., Cerri, C. E. P., de Oliveira, R. C., Durigan, M., Moura, N., Oliveira, V. H. F., Ribas, C., Vaz-de-Mello, F., Vieira, I., Zanetti, R., & Barlow, J. (2022). Linking land-use and land-cover transitions to their ecological impact in the Amazon. *Proceedings of the National Academy of Sciences, USA, 119*, e2202310119.

Öckinger, E., Schweiger, O., Crist, T. O., Debinski, D. M., Krauss, J., Kuussaari, M., Petersen, J. D., Pöyry, J., Settele, J., Summerville, K. S., & Bommarco, R. (2010). Life-history traits predict species responses to habitat area and isolation: a cross-continental synthesis: Habitat fragmentation and life-history traits. *Ecology Letters, 13*, 969–79.

O'Hara, R. B. (2005). Species richness estimators: how many species can dance on the head of a pin? *Journal of Animal Ecology, 74*, 375–86.

Otto, S. P. (2018). Adaptation, speciation and extinction in the Anthropocene. *Proceedings of the Royal Society B: Biological Sciences, 285*, 20182047.

Outhwaite, C. L., McCann, P., & Newbold, T. (2022). Agriculture and climate change are reshaping insect biodiversity worldwide. *Nature, 605*, 97–102.

Partridge, L., & Harvey, P. H. (1988). The ecological context of life history evolution. *Science, 241*, 1449–55.

Pescott, O. L., Simkin, J. M., August, T. A., Randle, Z., Dore, A. J., & Botham, M. S. (2015). Air pollution and its effects on lichens, bryophytes, and lichen-feeding Lepidoptera: Review and evidence from biological records. *Biological Journal of the Linnean Society, 115*, 611–35.

Pilotto, F., Rojas, A., & Buckland, P. I. (2022). Late Holocene anthropogenic landscape change in northwestern Europe impacted insect biodiversity as much as climate change did after the last Ice Age. *Proceedings of the Royal Society B: Biological Sciences, 289*, 20212734.

Pinkert, S., Barve, V., Guralnick, R., & Jetz, W. (2022). Global geographical and latitudinal variation in butterfly species richness captured through a comprehensive country-level occurrence database. *Global Ecology and Biogeography, 31*, 830–39.

Pontarp, M., Brännström, Å., & Petchey, O. L. (2019a). Inferring community assembly processes from macroscopic patterns using dynamic eco-evolutionary models and Approximate Bayesian Computation (ABC). *Methods in Ecology and Evolution, 10,* 450–60.

Pontarp, M., Bunnefeld, L., Cabral, J. S., Etienne, R. S., Fritz, S. A., Gillespie, R., Graham, C. H., Hagen, O., Hartig, F., Huang, S., Jansson, R., Maliet, O., Münkemüller, T., Pellissier, L., Rangel, T. F., Storch, D., Wiegand, T., & Hurlbert, A. H. (2019b). The latitudinal diversity gradient: Novel understanding through mechanistic eco-evolutionary models. *Trends in Ecology & Evolution, 34,* 211–23.

Powell, J. A. (2001). Longest insect dormancy: Yucca Moth larvae (Lepidoptera: Prodoxidae) metamorphose after 20, 25, and 30 years in diapause. *Annals of the Entomological Society of America, 94,* 677–80.

Pöyry, J., Carvalheiro, L. G., Heikkinen, R. K., Kühn, I., Kuussaari, M., Schweiger, O., Valtonen, A., van Bodegom, P. M., & Franzén, M. (2017). The effects of soil eutrophication propagate to higher trophic levels: Effects of soil eutrophication on herbivores. *Global Ecology and Biogeography, 26,* 18–30.

Pöyry, J., Paukkunen, J., Heliölä, J., & Kuussaari, M. (2009). Relative contributions of local and regional factors to species richness and total density of butterflies and moths in semi-natural grasslands. *Oecologia, 160,* 577–87.

Promislow, D. E. L., & Harvey, P. H. (1990). Living fast and dying young: A comparative analysis of life-history variation among mammals. *Journal of Zoology, 220,* 417–37.

Pyšek, P., Hulme, P. E., Simberloff, D., Bacher, S., Blackburn, T. M., Carlton, J. T., Dawson, W., Essl, F., Foxcroft, L. C., Genovesi, P., Jeschke, J. M., Kühn, I., Liebhold, A. M., Mandrak, N. E., Meyerson, L. A., Pauchard, A., Pergl, J., Roy, H. E., Seebens, H., Kleunen, M., Vilà, M., Wingfield, M. J., & Richardson, D. M. (2020). Scientists' warning on invasive alien species. *Biological Reviews, 95,* 1511–34.

Quinn, R. M., Gaston, K. J., Blackburn, T. M., & Eversham, B. (1997a). Abundance-range size relationships of macrolepidoptera in Britain: The effects of taxonomy and life history variables. *Ecological Entomology, 22,* 453–61.

Quinn, R. M., Gaston, K. J., & Roy, D. (1997b). Coincidence between consumer and host occurrence: Macrolepidoptera in Britain. *Ecological Entomology, 22,* 197–208.

Quintero, I., & Jetz, W. (2018). Global elevational diversity and diversification of birds. *Nature, 555,* 246–50.

Rabosky, D. L. (2013). Diversity-dependence, ecological speciation, and the role of competition in macroevolution. *Annual Review of Ecology, Evolution, and Systematics, 44,* 481–502.

Rabosky, D. L. (2021). Macroevolutionary thermodynamics: Temperature and the tempo of evolution in the tropics. *PLOS Biology, 19,* e3001368.

Rabosky, D. L., & Hurlbert, A. H. (2015). Species richness at continental scales is dominated by ecological limits. *The American Naturalist, 185,* 572–83.

Rees, M., Kelly, D., & Bjørnstad, O. N. (2002). Snow tussocks, chaos, and the evolution of mast seeding. *The American Naturalist, 160,* 44–59.

Regier, J. C., Mitter, C., Mitter, K., Cummings, M. P., Bazinet, A. L., Hallwachs, W., Janzen, D. H., & Zwick, A. (2017). Further progress on the phylogeny of Noctuoidea (Insecta: Lepidoptera) using an expanded gene sample. *Systematic Entomology, 42,* 82–93.

Reijenga, B. R., Murrell, D. J., & Pigot, A. L. (2021). Priority effects and the macroevolutionary dynamics of biodiversity. *Ecology Letters, 24,* 1455–66.

Remmel, T., Davison, J., & Tammaru, T. (2011). Quantifying predation on folivorous insect larvae: The perspective of life-history evolution. *Biological Journal of the Linnean Society, 104,* 1–18.

Román-Palacios, C., & Wiens, J. J. (2020). Recent responses to climate change reveal the drivers of species extinction and survival. *Proceedings of the National Academy of Sciences, USA, 117,* 4211–17.

Rönkä, K., Valkonen, J. K., Nokelainen, O., Rojas, B., Gordon, S., Burdfield-Steel, E., & Mappes, J. (2020). Geographic mosaic of selection by avian predators on hindwing warning colour in a polymorphic aposematic moth. *Ecology Letters, 23,* 1654–63.

Root, H. T., Verschuyl, J., Stokely, T., Hammond, P., Scherr, M. A., & Betts, M. G. (2017). Plant diversity enhances moth diversity in an intensive forest management experiment. *Ecological Applications, 27,* 134–42.

Roth, N., Hacker, H. H., Heidrich, L., Friess, N., García-Barros, E., Habel, J. C., Thorn, S., & Müller, J. (2021). Host specificity and species colouration mediate the regional decline of nocturnal moths in central European forests. *Ecography, 44,* 941–52.

Roy, K., Collins, A. G., Becker, B. J., Begovic, E., & Engle, J. M. (2003). Anthropogenic impacts and historical decline in body size of rocky intertidal gastropods in southern California. *Ecology Letters, 6,* 205–11.

Sabrosky, C. W. (1953). How many insects are there? *Systematic Zoology, 2,* 31–36.

Sæther, B.-E., Coulson, T., Grøtan, V., Engen, S., Altwegg, R., Armitage, K. B., Barbraud, C., Becker, P. H., Blumstein, D. T., Dobson, F. S., Festa-Bianchet, M., Gaillard, J.-M., Jenkins, A., Jones, C., Nicoll, M. A. C., Norris, K., Oli, M. K., Ozgul, A., & Weimerskirch, H. (2013). How life history influences population dynamics in fluctuating environments. *The American Naturalist, 182,* 743–59.

Saito, V. S., Perkins, D. M., & Kratina, P. (2021). A metabolic perspective of stochastic community assembly. *Trends in Ecology & Evolution, 36,* 280–83.

Satake, A., N. Bjørnstad, O., & Kobro, S. (2004). Masting and trophic cascades: Interplay between rowan trees, apple fruit moth, and their parasitoid in southern Norway. *Oikos, 104,* 540–50.

Satterfield, D. A., Sillett, T. S., Chapman, J. W., Altizer, S., & Marra, P. P. Seasonal insect migrations: Massive, influential, and overlooked. *Frontiers in Ecology and the Environment, 18,* 335–44.

Schauber, E. M., Ostfeld, R. S., & Evans, Jr., A. S. (2005). What is the best predictor of annual Lyme Disease incidence: Weather, mice, or acorns? *Ecological Applications, 15,* 575–86.

Schmidt, B. C., & Roland, J. (2006). Moth diversity in a fragmented habitat: Importance of functional groups and landscape scale in the boreal forest. *Annals of the Entomological Society of America, 99,* 1110–20.

Seddon, N., Merrill, R. M., & Tobias, J. A. (2008). Sexually selected traits predict patterns of species richness in a diverse clade of suboscine birds. *The American Naturalist, 171,* 620–31.

Seebens, H., Blackburn, T. M., Dyer, E. E., Genovesi, P., Hulme, P. E., Jeschke, J. M., Pagad, S., Pyšek, P., van Kleunen, M., Winter, M., Ansong, M., Arianoutsou, M.,

Bacher, S., Blasius, B., Brockerhoff, E. G., Brundu, G., Capinha, C., Causton, C. E., Celesti-Grapow, L., Dawson, W., Dullinger, S., Economo, E. P., Fuentes, N., Guénard, B., Jäger, H., Kartesz, J., Kenis, M., Kühn, I., Lenzner, B., Liebhold, A. M., Mosena, A., Moser, D., Nentwig, W., Nishino, M., Pearman, D., Pergl, J., Rabitsch, W., Rojas-Sandoval, J., Roques, A., Rorke, S., Rossinelli, S., Roy, H. E., Scalera, R., Schindler, S., Štajerová, K., Tokarska-Guzik, B., Walker, K., Ward, D. F., Yamanaka, T., & Essl, F. (2018). Global rise in emerging alien species results from increased accessibility of new source pools. *Proceedings of the National Academy of Sciences, USA, 115*, E2264–73.

Seifert, C. L., Strutzenberger, P., Hausmann, A., Fiedler, K., & Baselga, A. (2022). Dietary specialization mirrors Rapoport's rule in European geometrid moths. *Global Ecology and Biogeography, 31*, 1161–71.

Senior, V. L., Botham, M., & Evans, K. L. (2021). Experimental simulations of climate change induced mismatch in oak and larval development rates impact indicators of fitness in a declining woodland moth. *Oikos, 130*, 969–78.

Seymour, M., Brown, N., Carvalho, G. R., Wood, C., Goertz, S., Lo, N., & de Bruyn, M. (2020). Ecological community dynamics: 20 years of moth sampling reveals the importance of generalists for community stability. *Basic and Applied Ecology, 49*, 34–44.

Sharma, A., Kumar, V., Shahzad, B., Tanveer, M., Sidhu, G. P. S., Handa, N., Kohli, S. K., Yadav, P., Bali, A. S., Parihar, R. D., Dar, O. I., Singh, K., Jasrotia, S., Bakshi, P., Ramakrishnan, M., Kumar, S., Bhardwaj, R., & Thukral, A. K. (2019). Worldwide pesticide usage and its impacts on ecosystem. *SN Applied Sciences, 1*, 1446.

Shen, Z., Neil, T. R., Robert, D., Drinkwater, B. W., & Holderied, M. W. (2018). Biomechanics of a moth scale at ultrasonic frequencies. *Proceedings of the National Academy of Sciences, USA, 115*, 12200–5.

Simberloff, D. (1994). The ecology of extinction. *Acta Palaeontologica Polonica, 38*, 159–74.

Skogland, T. (1989). Natural selection of wild reindeer life history traits by food limitation and predation. *Oikos, 55*, 101–10.

Slade, E. M., Merckx, T., Riutta, T., Bebber, D. P., Redhead, D., Riordan, P., & Macdonald, D. W. (2013). Life-history traits and landscape characteristics predict macro-moth responses to forest fragmentation. *Ecology, 94*, 1519–30.

Southwood, T. R. E. (2021). Habitat, the templet for ecological strategies? *Journal of Animal Ecology, 46*, 336–65.

Spaak, J. W., Carpentier, C., & De Laender, F. (2021). Species richness increases fitness differences, but does not affect niche differences. *Ecology Letters, 24*, 2611–23.

Spitzer, K., Lepš, J., & Leps, J. (1988). Determinants of temporal variation in moth abundance. *Oikos, 53*, 31–36.

Stearns, S. C. (1989). Trade-offs in life-history evolution. *Functional Ecology, 3*, 259–68.

Stenseth, N. C., Falck, W., Bjornstad, O. N., & Krebs, C. J. (1997). Population regulation in snowshoe hare and Canadian lynx: asymmetric food web configurations between hare and lynx. *Proceedings of the National Academy of Sciences, USA, 94*, 5147–52.

Storch, D., Bohdalková, E., & Okie, J. (2018). The more-individuals hypothesis revisited: The role of community abundance in species richness regulation and the productivity-diversity relationship. *Ecology Letters, 21*, 920–37.

Strutzenberger, P., Brehm, G., Gottsberger, B., Bodner, F., Seifert, C. L., & Fiedler, K. (2017). Diversification rates, host plant shifts and an updated molecular phylogeny of Andean *Eois* moths (Lepidoptera: Geometridae). *PLOS ONE, 12*, e0188430.

Summerville, K. S., & Crist, T. O. (2004). Contrasting effects of habitat quantity and quality on moth communities in fragmented landscapes. *Ecography, 27*, 3–12.

Summerville, K. S., & Crist, T. O. (2003). Determinants of lepidopteran community composition and species diversity in eastern deciduous forests: Roles of season, eco-region and patch size. *Oikos, 100*, 134–48.

Summerville, K. S., & Crist, T. O. (2008). Structure and conservation of lepidopteran communities in managed forests of northeastern North America: A review. *The Canadian Entomologist, 140*, 475–94.

Svenningsen, C. S., Bowler, D. E., Hecker, S., Bladt, J., Grescho, V., Dam, N. M., Dauber, J., Eichenberg, D., Ejrnæs, R., Fløjgaard, C., Frenzel, M., Frøslev, T. G., Hansen, A. J., Heilmann-Clausen, J., Huang, Y., Larsen, J. C., Menger, J., Nayan, N. L. B. M., Pedersen, L. B., Richter, A., Dunn, R. R., Tøttrup, A. P., & Bonn, A. (2022). Flying insect biomass is negatively associated with urban cover in surrounding landscapes. *Diversity and Distributions, 28*, 1242–54.

Tallamy, D. W., Narango, D. L., & Mitchell, A. B. (2021). Do non-native plants contribute to insect declines? *Ecological Entomology, 46*, 729–42.

Tallamy, D. W., & Shriver, W. G. (2021). Are declines in insects and insectivorous birds related? *Ornithological Applications, 123*, duaa059.

Tammaru, T., Johansson, N. R., Õunap, E., & Davis, R. B. (2018). Day-flying moths are smaller: Evidence for ecological costs of being large. *Journal of Evolutionary Biology, 31*, 1400–4.

Tammaru, T., Ruohomäki, K., & Saloniemi, I. (1999). Within-season variability of pupal period in the autumnal moth: A bet-hedging strategy? *Ecology, 80*, 1666–77.

Thomas, C. D. (2015). Rapid acceleration of plant speciation during the Anthropocene. *Trends in Ecology & Evolution, 30*, 448–55.

Thomas, C. D., & Kunin, W. E. (1999). The spatial structure of populations. *Journal of Animal Ecology, 68*, 647–57.

Thompson, P. L., Guzman, L. M., De Meester, L., Horváth, Z., Ptacnik, R., Vanschoenwinkel, B., Viana, D. S., & Chase, J. M. (2020). A process-based metacommunity framework linking local and regional scale community ecology. *Ecology Letters, 23*, 1314–29.

Thomsen, P. F., Jørgensen, P. S., Bruun, H. H., Pedersen, J., Riis-Nielsen, T., Jonko, K., Słowińska, I., Rahbek, C., & Karsholt, O. (2016). Resource specialists lead local insect community turnover associated with temperature: Analysis of an 18-year full-seasonal record of moths and beetles. *Journal of Animal Ecology, 85*, 251–61.

Tielens, E. K., Cimprich, P. M., Clark, B. A., DiPilla, A. M., Kelly, J. F., Mirkovic, D., Strand, A. I., Zhai, M., & Stepanian, P. M. (2021). Nocturnal city lighting elicits a macroscale response from an insect outbreak population. *Biology Letters, 17*, 20200808.

Tilman, D. (2004). Niche tradeoffs, neutrality, and community structure: A stochastic theory of resource competition, invasion, and community assembly. *Proceedings of the National Academy of Sciences, USA, 101*, 10854–61.

Trisos, C. H., Merow, C., & Pigot, A. L. (2020). The projected timing of abrupt ecological disruption from climate change. *Nature, 580,* 496–501.

Udy, K., Fritsch, M., Meyer, K. M., Grass, I., Hanß, S., Hartig, F., Kneib, T., Kreft, H., Kukunda, C. B., Pe'er, G., Reininghaus, H., Tietjen, B., Tscharntke, T., Waveren, C., & Wiegand, K. (2021). Environmental heterogeneity predicts global species richness patterns better than area. *Global Ecology and Biogeography, 30,* 842–51.

Uhl, B., Wölfling, M., & Fiedler, K. (2021). From forest to fragment: compositional differences inside coastal forest moth assemblages and their environmental correlates. *Oecologia, 195,* 453–67.

Usher, M. B., & Keiller, S. W. J. (1998). The macrolepidoptera of farm woodlands: Determinants of diversity and community structure. *Biodiversity and Conservation, 7,* 725–48.

Valtonen, A., Hirka, A., Szőcs, L., Ayres, M. P., Roininen, H., & Csóka, G. (2017). Long-term species loss and homogenization of moth communities in Central Europe. *Journal of Animal Ecology, 86,* 730–38.

van Klink, R., August, T., Bas, Y., Bodesheim, P., Bonn, A., Fossøy, F., Høye, T. T., Jongejans, E., Menz, M. H. M., Miraldo, A., Roslin, T., Roy, H. E., Ruczyński, I., Schigel, D., Schäffler, L., Sheard, J. K., Svenningsen, C., Tschan, G. F., Wäldchen, J., Zizka, V. M. A., Åström, J., & Bowler, D. E. (2022). Emerging technologies revolutionise insect ecology and monitoring. *Trends in Ecology & Evolution,* S0169534722001343.

Van Nieukerken, E. J., Kaila, L., Kitching, I. J., Kristensen, N. P., Lees, D. C., Minet, J., Mitter, C., Mutanen, M., Regier, J. C., Simonsen, T. J., Wahlberg, N., Yen, S.-H., Zahiri, R., Adamski, D., Baixeras, J., Bartsch, D., Bengtsson, B. Å., Brown, J. W., Bucheli, S. R., Davis, D. R., Prins, J. D., Prins, W. D., Epstein, M. E., Gentili-Poole, P., Gielis, C., Hättenschwiler, P., Hausmann, A., Holloway, J. D., Kallies, A., Karsholt, O., Kawahara, A. Y., Koster, S. J. C., Kozlov, M. V., Lafontaine, J. D., Lamas, G., Landry, J.-F., Lee, S., Nuss, M., Park, K.-T., Penz, C., Rota, J., Schintlmeister, A., Schmidt, B. C., Sohn, J.-C., Solis, M. A., Tarmann, G. M., Warren, A. D., Weller, S., Yakovlev, R. V., Zolotuhin, V. V., & Zwick, A. (2011). Order Lepidoptera Linnaeus, 1758. In Z.-Q. Zhang (Ed.). Animal biodiversity: An outline of higher-level classification and survey of taxonomic richness. *Zootaxa, 3148,* 212.

Varley, G. C., & Gradwell, G. R. (1960). Key factors in population studies. *Journal of Animal Ecology, 29,* 399.

Varley, G. C., & Gradwell, G. R. (1962). The effect of partial defoliation by caterpillars on the timber production of oak trees in England. *Proceedings of the XI International Congress of Entomology, Wien,* 211–14.

Volf, M., Volfová, T., Seifert, C.L., Ludwig, A., Engelmann, R. A., Jorge, L. R., Richter, R., Schedl, A., Weinhold, A., Wirth, C., & van Dam, N. M. (2022). A mosaic of induced and non-induced branches promotes variation in leaf traits, predation and insect herbivore assemblages in canopy trees. *Ecology Letters, 25,* 729–39.

Wagner, D. L., Fox, R., Salcido, D. M., & Dyer, L. A. (2021). A window to the world of global insect declines: Moth biodiversity trends are complex and heterogeneous. *Proceedings of the National Academy of Sciences, USA, 118,* e2002549117.

Wagner, D. L., & Van Driesche, R. G. (2010). Threats posed to rare or endangered insects by invasions of nonnative species. *Annual Review of Entomology, 55,* 547–68.

Wahlberg, N., Wheat, C. W., & Peña, C. (2013). Timing and patterns in the taxonomic diversification of Lepidoptera (butterflies and moths). *PLoS ONE, 8*, e80875.

West, C. (1985). Factors underlying the late seasonal appearance of the lepidopterous leaf-mining guild on oak. *Ecological Entomology, 10*, 111–20.

Wheat, C. W., Vogel, H., Wittstock, U., Braby, M. F., Underwood, D., & Mitchell-Olds, T. (2007). The genetic basis of a plant insect coevolutionary key innovation. *Proceedings of the National Academy of Sciences, 104*, 20427–31.

Whitaker, M. R. L., & Salzman, S. (2020). Ecology and evolution of cycad-feeding Lepidoptera. *Ecology Letters, 23*, 1862–77.

Wickman, P.-O., & Karlsson, B. (1989). Abdomen size, body size and the reproductive effort of insects. *Oikos, 56*, 209–14.

Wiens, J. J. (2021). Vast (but avoidable) underestimation of global biodiversity. *PLOS Biology, 19*, e3001192.

Wiens, J. J., & Donoghue, M. J. (2004). Historical biogeography, ecology and species richness. *Trends in Ecology & Evolution, 19*, 639–44.

Wilson, J. F., Baker, D., Cheney, J., Cook, M., Ellis, M., Freestone, R., Gardner, D., Geen, G., Hemming, R., Hodgers, D., Howarth, S., Jupp, A., Lowe, N., Orridge, S., Shaw, M., Smith, B., Turner, A., & Young, H. (2018). A role for artificial night-time lighting in long-term changes in populations of 100 widespread macro-moths in UK and Ireland: A citizen-science study. *Journal of Insect Conservation, 22*, 189–96.

Wilson, J. F., Baker, D., Cook, M., Davis, G., Freestone, R., Gardner, D., Grundy, D., Lowe, N., Orridge, S., & Young, H. (2015). Climate association with fluctuation in annual abundance of fifty widely distributed moths in England and Wales: A citizen-science study. *Journal of Insect Conservation, 19*, 935–46.

Winkler, I. S., & Mitter, C. (2008). The phylogenetic dimension of insect-plant interactions: A review of recent evidence. In K. J. Tilmon (Ed.). *Specialization, speciation and radiation: The evolutionary biology of herbivorous insects* (pp. 240–63). University of California Press.

Wood, T. J., & Goulson, D. (2017). The environmental risks of neonicotinoid pesticides: A review of the evidence post 2013. *Environmental Science and Pollution Research, 24*, 17285–325.

Yiukawa, J. (1986). Moths collected from the Krakatau Islands and Panaitan Island, Indonesia. *Tyo to Ga, 36*, 181–84.

Yoneda, M., & Wright, P. (2004). Temporal and spatial variation in reproductive investment of Atlantic cod *Gadus morhua* in the northern North Sea and Scottish west coast. *Marine Ecology Progress Series, 276*, 237–48.

Zeuss, D., Brunzel, S., & Brandl, R. (2017). Environmental drivers of voltinism and body size in insect assemblages across Europe: Voltinism and body size in insect assemblages. *Global Ecology and Biogeography, 26*, 154–65.

Notes

Introduction

1. See: Townsend, C. R., Begon, M., & Harper, J. L. (2003). *Essentials of Ecology* (2nd ed.). Blackwell Publishing.

2. Andrewartha, H. G. (1961). *Introduction to the Study of Animal Populations*. Methuen.

3. Krebs, C. J. (1972). *Ecology*. Harper & Row.

4. Darwin, C. (1859). *On the Origin of Species by Means of Natural Selection, or, the Preservation of Favoured Races in the Struggle for Life*. J. Murray.

Chapter 1

1. Forbush, E. H., & Fernald, C. H. (1896). *The Gypsy Moth*. PORTHETRIA DIS-PAR *(LINN.)*. *A Report of the Work of destroying the insect in the commonwealth of Massachusetts, together with an Account of its History and Habits both in Massachusetts and Europe*. Wright & Potter Printing Co.

2. See: http://www.dcscience.net/2020/03/23/exponential-growth-is-terrifying.

Chapter 2

1. Manley, C. (2015). *British Moths* (2nd ed.). Bloomsbury.

Chapter 5

1. UK Moth Recorders Meeting, January 30, 2021, https://www.youtube.com/watch?v=8yRPZdVKs5g.

Chapter 6

1. Abang, F., & Karim, C. (2005) Diversity of macromoths (Lepidoptera: Heterocera) in the Poring Hill Dipterocarp Forest, Sabah, Borneo. *Journal of Asia-Pacific Entomology, 8*, 69–79.

Chapter 7

1. Lowen, J. (2021) *Much Ado about Mothing: A year intoxicated by Britain's rare and remarkable moths*. Bloomsbury Wildlife.

Chapter 8

1. WWF (2020) *Living Planet Report 2020: Bending the curve of biodiversity loss*. Almond, R. E. A., Grooten M., & Petersen, T. (eds). World Wildlife Foundation.

2. Wood, T. J., & Goulson, D. (2017). The environmental risks of neonicotinoid pesticides: A review of the evidence post 2013. *Environmental Science & Pollution Research, 24*, 17285–325.

3. Jepson, P. D., et al. (2016). PCB pollution continues to impact populations of orcas and other dolphins in European waters. *Scientific Reports, 6*, 18573.

4. See: https://www.fao.org/sustainability/news/detail/en/c/1274219.

Acknowledgments

This book would never have happened if Will Francis and Claire Conrad at Janklow & Nesbit had not believed that I could write it, and passed that belief on to me. Basing the book around the moth trap was Claire's suggestion—a lightbulb moment. Will Francis and Ian Bonaparte then found publishers prepared to take a punt on a book weaving the science of ecology with the natural history of moths. Without them, I'd never have had the great fortune to work with two fantastic editors in Jenny Lord at Weidenfeld & Nicolson, and Rebecca Bright at Island Press. Jenny was instrumental in nursing my brittle self-confidence through the difficult process of actually doing the writing, commenting on drafts of the early chapters and then the first draft of the whole book. Rebecca's sharp eye added extra polish, on top of adapting the text for an American audience. Further shine was added by my astute copy editors, Francine Brody and Mike Fleming. I could not have asked for better support through the whole production process—thank you all!

I'm quite sure that this book will annoy ecologists and Aurelians alike. In my defense, I would say that ecology and moths are both vast areas of knowledge, and that one person will never be able to do justice to either field in one short book. I hope that at least the ecologists enjoy the bits on moths, and the moth-ers the bits on ecology. As it goes, I've had fantastic support over the years from experts in both communities, who have been unfailingly generous with their knowledge and time and made this book possible. So really, you only have yourselves to blame.

In the world of moths, thank you to Sean Foote of @MothIDUK and Tom August of What's Flying Tonight for their incredible online identification resources, and Phil Barden, Barry Henwood, Richard Lewington, James Lowen, Colin Plant, Ben Sheldon, and Chris Wilkinson for helping me to put names to moths. I wish I could tell Douglas Boyes how much I learnt from him. His early death was a tragic loss to the moth and science communities. I cannot imagine how difficult it must be for his family and friends—my heart goes out to them.

I've talked ecology over the years with more people than I could possibly list here, and I'd like to thank them all for taking the time to share their expertise. Some of those ecologists have been particularly influential in the evolution of my own views, and special thanks go to Phill Cassey, Jane Catford, Steven Chown, Ben Collen, Richard Duncan, Ellie Dyer, Kevin Gaston, Charles Godfray, Andy Gonzalez, Richard Gregory, Paul Harvey, Bob Holt, Kate Jones, John Lawton, Julie Lockwood, Georgina Mace, Tim Newbold, Ian Owens, Alex Pigot, Stuart Pimm, Petr Pysek, Dave Richardson, and Helen Roy. I sorely miss Ben and Georgina.

Gavin Broad, Chris Raper, and Seirian Sumner helped answer my questions about wasps. Emma Milnes and Ann Sylph facilitated access to the fantastic collections at the ZSL Library. Jon Bridle and Seirian Sumner provided key pieces of inspiration during the writing process. Over the years, the Universities of Manchester, Oxford, Birmingham, Lincoln (New Zealand), and Adelaide, as well as Imperial College, the Institute of Zoology, and UCL have all provided supportive environments in which to pursue my study and research. No institution is perfect, but it has been a pleasure to work in academia.

I am very grateful to Joanna Blackburn, Sam Fanaken, Charlie Outhwaite, Helen Roy, Ben Sheldon, and Susie Wesson for reading and commenting on a complete draft of the text. Their feedback was most helpful in polishing the final version. Any errors and misrepresentations that remain are entirely my own.

I owe the largest debt of gratitude to my family. They have been unfailingly supportive in all my life choices, and I'm sorry that their long wait for me to get a proper job continues. Mum, Dad, Louise, John, Joanna, Barnaby, and Hugo—thank you all. And last but not least, thank you Noëlle and Milly, the two most beautiful links in the unbroken chain of life.

Index

Page numbers followed by "f" indicate images. Page numbers followed by "n" indicate footnotes.

About the Author

Tim Blackburn is a scientist with thirty years' experience studying questions about the distribution, abundance, and diversity of species in ecological assemblages. He is currently Professor of Invasion Biology at University College London, where his research focuses on alien species and his teaching mainly involves leading field courses. Before that he was the Director of the Institute of Zoology, the research arm of the Zoological Society of London.